Big Data Management and Analytics

As more companies go digital and conduct their business online, this book provides practical examples of how they can better manage their data and use it to generate maximum value. It offers an integrated approach by treating data as an asset and discusses how to preserve and protect it just like any other corporate asset.

Big Data Management and Analytics: Concepts, Tools, and Applications illustrate effective strategies for managing, governing, and analyzing big data to gain a competitive edge for companies utilizing big data and analytics. It offers a comprehensive guide on methods, tools, and concepts to efficiently manage and analyze big data in order to make informed decisions. Additionally, this book explores the significance of artificial intelligence and machine learning in leveraging big data and how they can be optimized in a well-structured environment. This book also emphasizes treating big data as a valuable asset and outlines strategies for preserving and safeguarding it like any other corporate asset. The inclusion of case studies ensures that the methodologies and concepts presented can be easily implemented in day-to-day operations.

Given the current significance of big data in the business world, this book equips readers with the necessary skills to effectively manage this valuable asset. It is tailored for practitioners, students, and professionals working in data mining, big data, and machine learning across various industries, including manufacturing.

Big Data Management and Analytics
Concepts, Tools, and Applications

Rajesh Jugulum, David J. Fogarty,
Chris Heien, and Surya Putchala

CRC Press
Taylor & Francis Group
Boca Raton London New York

CRC Press is an imprint of the
Taylor & Francis Group, an **informa** business

Designed cover image: Shutterstock

First edition published 2025
by CRC Press
2385 NW Executive Center Drive, Suite 320, Boca Raton FL 33431

and by CRC Press
4 Park Square, Milton Park, Abingdon, Oxon, OX14 4RN

CRC Press is an imprint of Taylor & Francis Group, LLC

ISBN: 978-1-032-04040-0 (hbk)
ISBN: 978-1-032-04043-1 (pbk)
ISBN: 978-1-003-19032-5 (ebk)

DOI: 10.1201/9781003190325

Typeset in Times
by codeMantra

Contents

Preface and Acknowledgments

The authors were inspired to write this book after working for several large firms where data was not being treated as an important corporate asset like other assets on the firm's balance sheet. After realizing the importance of data, organizations need to view data as a critical asset, equally important as other assets such as people, capital, raw materials, and facilities. The concept of data as an asset has driven the need for dedicated data management programs, and in some cases, the appointment of a chief data officer. The companies we have worked with have all types of data (which we typically call big data) – numerical, text, social media, mobile applications, etc. It is very important to manage and analyze big data with a disciplined approach, tools, and concepts. This involves developing a sound big data strategy and using appropriate concepts, tools, methods, and approaches for the effective execution of analytics to help differentiate the organization from competitors. One of the important strategic aspects should be to have a process in place to ensure data quality, so the data is fit for its purpose. Another important strategic topic is data valuation and understanding exactly what a firm's data is worth. How can we expect firms to treat data as an asset if they don't have a good handle on what their data is worth?

This book aims to provide an integrated approach for big data management and analytics. It offers an overview of various big data-related concepts, tools, methods, and approaches. These include approaches for data quality, analytics execution, cloud-based analytics, and AI and Gen AI-related topics. The set of topics covered in this book is unique and guarantees the development of a robust big data strategy with a focus on ensuring high-quality data as the raw material for analytics, so that appropriate business decisions can be made. The case studies presented will allow users to apply these methods in different real-life situations while understanding the methodology.

Although many companies are using big data in their operations, there are not many books that deal with big data management and analytics. This book will close that gap by providing methods, tools, and concepts to manage big data and analyze it to make informed data decisions. It also describes how artificial intelligence and machine learning methods depend on the presence of big data and can be more effective in a better-managed big data environment.

This book is designed for managers and leaders who will have the management of big data as part of their span of control. It treats big data as an asset and informs managers on how to value, preserve, and protect the asset just like any other corporate asset. This book broadens the aspects of big data management, governance, strategy, quality, and analytics, so data managers can be more effective and bring a new perspective.

Finally, this book will demonstrate to readers how to manage, govern, value, and analyze big data and, as such, develop a competitive advantage for their firms through the use of big data and analytics.

We hope that readers of this book will develop a heightened appreciation for how big data is an important and valuable asset for their firms and that by learning and using the concepts, tools, methods, and approaches for the effective execution of big data and analytics, they will set their firms on a new trajectory to compete in the marketplace through more effective data-driven decisions and operations.

We are very grateful to CRC Press for giving us the opportunity to publish this book. We are particularly thankful to Cindy Carelli, Executive Editor, for her help and support and for being flexible in accommodating many requests from us. Thanks are also due to Kaitlyn Fisher, Editorial Assistant, for her cooperation during this project. We would like to thank Uma Maheswari of Codemantra for the help and support she provided. Finally, we are thankful to Wiley, ASQ Quality Press, and CRC Press for allowing me to use parts of previous publications.

Thanks are due to Brian Bramson, Bob Granese, Chuan Shi, Raji Ramachandran, Ian Joyce, Jagmeet Singh, Laura Sebastian-Coleman, and Raj Vadlamudi for their involvement and help in putting together data quality approach and conducting case studies that are presented in Chapter 4.

We also would like to thank Anush Jugulum for carefully reading some chapters of the manuscript and making changes.

Finally, our special thanks to all the individuals who relentlessly contributed/are contributing to the field of data.

Authors

Rajesh Jugulum, PhD, is a data science innovation, analytics and process engineering leader. He teaches Data Science Related Courses at Northeastern University in Boston. He has experience in these areas in different types of industries including healthcare, finance, and manufacturing. He held executive positions in the areas of process engineering and data science at Citi Group and Bank of America. Before joining the financial industry, Rajesh was at Massachusetts Institute of Technology where he was involved in research and teaching.

Rajesh is the author/co-author of several papers and six books including books on common data sense, data quality, and design for six sigma. He is an American Society for Quality (ASQ) Fellow and his other honors include ASQ's Feigenbaum Medal, International Technology Institute's Rockwell Medal, and a Recognition Award from Industrial and Systems Engineering Department of Wayne State University. He has been listed in the "Who's Who in the World" list by Marquis Who's Who publication board. He was featured as "Face of Quality" in an issue of Quality Progress magazine and his profile was also published in the *Journal of Quality Engineering Society*. He has delivered talks as the keynote speaker at several conferences, symposiums, and events related to data analytics and process engineering. He has also delivered lectures at several universities/companies across the globe and participated as a judge in data-related competitions.

David Fogarty, PhD, MBA, currently works as a part-time professor of business management at National University. Prior to this he worked for one of the largest global health insurers as their Chief Marketing Analytics Officer and Head of Global Customer Value Management and Analytics.

For 20 years, David also worked at the General Electric Company and has held quantitative analysis leadership roles in the various business units of the company across several functions including risk management and marketing both internationally and in the US. He has over 15 US patents or patents pending on business analytics algorithms and is a certified Six Sigma Master Black Belt in Quality which is the highest qualification within the Six Sigma Quality methodology.

David has over 20 years of teaching experience having held various adjunct academic appointments at both the graduate and undergraduate levels in statistics, international management, and quantitative analysis. He has also taught business analytics courses at the esteemed GE Crotonville Management Development Institute in Crotonville, New York. He has 50 published research papers in peer-reviewed academic journals and has also published five books. His research interests include how to conduct analysis with missing data, the cultural meaning of data, integrating machine learning and artificial intelligence algorithms into the statistical science framework and many other topics related to quantitative analysis in business.

Chris Heien, PhD, was one of the inaugural graduates of the Information Quality in Applied Sciences doctoral program, the first of its kind in the world, from the University of Arkansas at Little Rock in 2012. During this time he taught three years of object-oriented software engineering and presented his research in data supply chains across conferences in North America, South America, and Asia. His research has continued through the co-founding of a software company in 2009 specializing in the analysis of Information Manufacturing Systems through Information Product Maps, through which he has consulted on cloud architecture strategies for early stage startups in Texas, New York, and Massachusetts.

In the past 13 years, Chris has held several roles as a practitioner in the finance and insurance industries including Cigna, GE Capital, and Citigroup. In these roles, he applied his background in software and cloud architecture toward various initiatives such as developing data quality monitoring strategies, guiding model execution in big data environments, data science model delivery in continuous integration, continuous delivery (CICD) pipelines, and implementing anomaly detection monitoring within high-risk, high-cost scenarios.

Surya Putchala is a Technology leader with expertise in driving measurable operational improvements for Fortune 500 companies and startups through innovative consulting and transformation solutions. As an entrepreneur and principal, Surya has spearheaded the development of three innovative AI solutions, resulting in $20 million in mergers and acquisitions. His expertise spans Generative AI, Large Language Models (LLMs), Data Science, and Machine Learning, with a strong focus on delivering tangible business outcomes.

Throughout his career, Surya has demonstrated his ability by building high-performing teams and establishing Centers of Excellence. He has mentored over 300 employees and guided six startups, contributing to the advancement of AI-driven innovations.

Surya has delivered over 50 Proofs of Concept across finance, retail, and healthcare industries, generating significant revenues and acquiring marquee clients. He has been named a LinkedIn Top Voice in AI and actively contributes to the AI community through mentoring, publishing, and speaking engagements. He combines technical acumen with strategic insights, making him a trusted advisor for organizations aiming to harness the power of AI and data-driven strategies.

1 The Management of BIG Data Overview

CHAPTER 1 - BIG DATA FUNDAMENTALS: SMALL DATA, BIG DATA, RELEVANT DATA

- Overview of the course and expectations of the book.
- Big Data history
- Review of the high-level concepts of where, what, why, and how Big Data and Business converge
- Big Data 5 Vs – volume, velocity, variety, veracity, and value
- The promise and challenges of Big Data
- Data for All

1.1 HISTORY OF BIG DATA

As companies collect ever more data, perpetuated by the use of technology and e-commerce, many business leaders are under the false belief that Big Data is a new phenomenon. In reality, paying attention to Big Data is a new phenomenon for organizations, but many forms of data are already familiar with Big Data. For example, when we just talk about data itself, interestingly, it precedes the modern computer by over 7,000 years and coincides with the birth of Western Civilization! The first recorded data was crop and livestock inventories gathered in ancient Mesopotamia. Ancient Mesopotamia is hypothesized to be where modern agriculture began, and the first data collected was of a bookkeeping nature, recording agricultural stocks, offering rations, and taxes on clay tablets.

The next major event recorded in data science occurred after the birth of Christianity, when in 1663, John Gaunt created the first mortality tables using death records in England. Gaunt is considered to be the father of modern statistics. With Gaunt's mortality tables, people could now start seeing trends in life expectancy, which opened the door for the concept of life insurance to be created. Sir Edmund Halley of was one of the first people recognized for using this information in such a manner, although perhaps he is more famous for his discovery of the famous comet, aptly named after him.

For the next major stepping stone to modern Big Data, we move to the mid-20th century when governments started collecting data on large mainframe computers for tax, defense, and social welfare program administration. Computers at this time were reserved for large organizations, and access to them was only through programmers with knowledge of programming languages, such as COBOL or Assembler. In the middle of all this, Dr. Grace Hopper discussed the real value of computers is

DOI: 10.1201/9781003190325-1

evaluating the way information flows through an organization. Moreover, Bill Inmon created the concept of the data warehouse, which was the first time data was systematically separated from operational systems just for analysis and reporting purposes. Focus forward to the personal computer and companies such as Apple, Dell, and Microsoft, where computing technology became available to a large portion of the population.

In 1989, British computer scientist Tim Berners-Lee invented what eventually came to be known as the World Wide Web. Simply desiring to share information via a "hypertext" system, he had no idea at the time of the profound impact of his invention would have, which would eventually transform the world in terms of access to information. As the World Wide Web became more embedded in personal and professional activities and e-commerce grew, so did the data that organizations captured.

The term Big Data term was coined in 2005 by Roger Mouglas from O'Reilly Media. They also created the term "Web 2.0". Of course, in the age of the internet and Big Data, the reign of SQL and relational databases was limited due to the huge amounts of unstructured data. Hadoop was launched in 2005 by Yahoo and built on top of Google's MapReduce. It is currently used by a large number of firms, as opposed to relational databases, to store and crunch Big Data.

Web 2.0 kicked in next with social media sites, and with the rise of mobile apps, the amount of data produced and stored expanded exponentially. The concept of SOLOMO was coined for service businesses, which stood for Social, Local, Online, and Mobile. For manufacturing, the term IOT (Internet of Things) was coined. The IOT connected any device with an on/off switch via the internet. One couldn't even begin to imagine how much data was collected with such online innovations.

This represents the timeline and events leading up to what we recognize today when we hear the term Big Data.

1.2 WHERE, WHAT, WHY, AND HOW BIG DATA AND BUSINESS CONVERGE

1.2.1 What Exactly Is Big Data? Big Data Defined

The first definition of Big Data we will explore is that it is a very large dataset, almost impossible to manage and analyze using traditional data mining tools like Excel, SPSS, or even the previously best-in-class scalable platform SAS. Of course, this very large dataset contains large volumes of data – both structured and unstructured – that inundates a business on a day-to-day basis. When this author set up the first Data Management Platform (DMP) at a large insurance company, we had only small B2B campaigns of maybe 30,000 business contacts receiving digital ads. However, the log files we received from the digital ads, which were being stored in our DMP, were counted in the millions. Of course, Big Data is more than just large amounts of data in that it also represents a new attitude by businesses, non-profits, government agencies, and individuals that combining data from multiple sources could lead to better decisions. It's a recognition that data is an asset similar to any "hard" corporate asset, such as factories and buildings. It must be noted that even though data is currently being recognized as a corporate asset, accountants are still developing the valuation techniques to compare these to hard assets. The stock market has a difficult time valuating these assets and, to much surprise, often overvalues them. Take the valuation of companies like Facebook, Netflix, and Twitter compared to General Electric or General Motors, both of which have many hard assets, including factories and equipment.

Another common definition of Big Data is that it consists of "extensive datasets – primarily in the characteristics of volume, velocity, variety and/or variability, and veracity – that require a scalable architecture for efficient storage, manipulation, and analysis". Volume represents the scale of data. Velocity represents the analysis of streaming data. Variety represents the different types of data, and Veracity represents the uncertainty of data. The four V's have now been extended to include Value, which represents the ability to achieve greater value through superior analytics. Over time, some additional Vs have been added to the definition of Big Data, which include Validity, which represents data quality, governance, and master data management on massive datasets; Variability, which represents dynamic evolving behavior in data source; Venue, which represents distributed, heterogeneous data from multiple platforms; Vocabulary, which represents data models and semantics that describes data structure; and Vagueness, which represents confusion over the meaning of Big Data and the tools used. I am sure data scientists are thinking up new V-words to describe Big Data. For example, one of the authors of this book introduced a new V-term for Big Data know as Vexing. Vexing represents the ability of Big Data to solve some of the world's most vexing problems. These problems range from healthcare to the environment. The 2020 Coronavirus outbreak was a direct example of how Big Data was used to help slow the spread of the pandemic and allow time for our healthcare resources to prepare for the outbreak and not become overwhelmed. Big Data promises to offer new solutions to these most difficult problems.

Finally, in addition to the V-words, some have defined Big Data as an amount of data that exceeds a **petabyte** – 1 million gigabytes. Just to provide some perspective on a petabyte, it represents 2 to the 50th power of bytes. This is equivalent to roughly 61,000 desktops or 170 top-end servers to add up to a single petabyte of RAM. A typical DVD holds 4.7 GB of data and one movie. A petabyte storage can hold 223,101 DVD-quality movies. There are some fun facts outlined by Lifewire below:

- To store a single PB would take over **745 million floppy disks** or **1.5 million CD-ROM discs**, clearly not an efficient way to collect a petabyte of information, but it's fun to think about!
- The movie Avatar needed **about 1 PB of storage** to render those graphics.
- It's estimated that **the human brain can store around 2.5 PB of memory data**.
- Over **3.4 years of 24/7 Full HD video recording** would be around 1 PB in size.
- As of late 2018, the Wayback Machine was **storing over 25 PB of data**!
- 1 PB is equivalent to **over 4,000 digital photos per day** over your entire life.

A single PB is 1,024 TB...you know, that number we already established was enormous even at one! In yet a more impressive view, 1 PB is equal to **over 1 quadrillion bytes**!

1.3 PROMISES AND CHALLENGES OF BIG DATA

In this section, we will examine both the promises and challenges of Big Data. Firstly, 75% of companies recently surveyed indicated that they would increase their investments in Big Data next year. Moreover, 58% of those same companies reported that data management is embedded throughout their business operations, and 95% of them also indicated that they didn't consider their business analysts to be a part of their technology staff. The latter is an interesting finding, since it is an indication that analytics is being considered a more strategic function and not being lumped into the same technology efforts required to "keep the lights on" at firms. Of course, not all technology staff should be considered less strategic. However, at many firms, this is the case, so a perception that analysts are part of the business is a step in the right direction toward companies really being able to capitalize on their Big Data investments. The dilemma in the world of Big Data, however, is that these two worlds cannot stand apart in the organization but must instead partner in order to fully realize the benefits of Big Data. When you have a requirement like this, there tends to be a discussion around consolidation. However, organizations are better served if the two teams can operate in a very tight partnership rather than consolidated, where a technology group with a bigger purpose is able to impose its own corporate culture on the analytics group,

Other findings include the fact that 73% of firms have leveraged data to increase revenues. Moreover, 43% have used data to locate new sources of revenue, and 57% have used data to increase an existing revenue stream. A full 84% of these executive attribute Big Data to helping them make better business decisions.

When asked if more work needed to be done, 85% reported obstacles in managing and analyzing their data. Some of the key trends causing companies to rethink their data management strategy include the increasing use of social media, mobile, and cloud technologies. These technologies not only change the economics of storing and analyzing data but also make it even more imperative for companies to focus on building their capabilities in using data analysis to drive business decisions.

In terms of emerging opportunities with Big Data, as data continues to grow exponentially in firms a little under half of firms are planning to invest in Big Data initiatives in the future and organizations are averaging an investment of $8-10 million per annum on Big Data initiatives. All of these estimates were developed prior to COVID-19. COVID-19 has taught us that Big Data is needed for making the most optimal decisions. The Executive branches of state and local governments openly discussed how they were using Big Data to make decisions and even displayed visualizations of their findings on national TV. People were also demanding data and quantitative answers related to the pandemic in order to justify and rationalize their compliance with government recommendations. During the pandemic, people wanted data to know whether it was safe to go to the grocery store or get takeout from restaurants. People also wanted to know whether it was safe to go on vacation to a certain place and what they had to do to protect themselves once they arrived. What was the infection rate at their destination? How could they avoid infection along the journey? What were the rules at the destination that vacationers had to follow? Moreover, companies wanted to know whether it was safe for their employees to return to work. Universities wanted to know whether a student should participate in classroom sessions. What had the students done in the past week that would put them at risk of entering the classroom? The same goes for the instructors. Therefore, it can be expected that more companies will be investing in Big Data as we continue to distance ourselves from the crisis. We are already seeing this in terms of the growth of cloud computing, although I suspect that due to work-at-home requirements, more companies, especially those facing increased attrition after return-to-work policies post-COVID-19, will outsource more activities and infrastructure to the cloud in order to cope with reduced density in the workplace. There is now a new justification for not having everything on-premise due to the reduced population density in the workplace. IT organizations were among the most densely staffed firms in a large organization pre-COVID. In one large health services-related entity with which the authors are affiliated, this is highlighted by a basement that represented a former Cold War fallout shelter packed with resources from the IT department, mostly from India or other offshore locations with working vias from the US. This company is literally running out of space to contain its IT workers, and the subsequent conditions in which they are working are not at all conducive to social distancing nor is the workspace anything representing a professional Fortune 1000 working environment. Now, there will be employs leaving strict return-to-work policies for hybrid work environments. This will also speed up the transition to cloud computing and AI automation. I suppose offshoring will also become more popular as companies try to compensate for the aforementioned attrition they will face. However, while compensating for the lack of employees at home through offshoring, this could create risks in the service supply chain. These risks can come about in the form of susceptibility due to natural disasters and political and military conflicts. These risks can be mitigated in different ways, such as through multiple offshore centers in addition to nearshoring. However, one thing is for certain: the movement of Big Data into the cloud environment, which will facilitate unlimited storage and processing power, will also make access to data easier for offshore labor, and further advancing the migration of critical IT and analytics resources offshore. All in all, Big Data and the analytics associated with it are now embedded into our personal and professional lives to the extent that any changes in the world will now be addressed through the analysis of Big Data. In other words, more and more solutions will be created from analytics moving forward, rather than human judgment. Human judgments will prove less reliable and will be subject to more scrutiny if the data is not there to back them up. This will continue to increase the demand for Big Data solutions and the analytics required to derive insights from all of this data. It is predicted that a large portion of the economy will be devoted to finding and implementing improvements in how the world operates through insights derived from the data. This will be a positive for those seeking careers in data science and will bring to life the concept of a data economy, where data and insights are traded like physical assets. As someone who has been in the analytics field for over 30 years, it's an interesting perspective to note that Big Data has become even bigger in terms of global impact than anyone could have ever imagined. It has permeated every aspect of our lives and drives decisions across a multitude of fields, ranging from how we get out power delivered across the grid to whether a package is going to arrive on time.

Statistics show that currently over 50% of organizations are either implementing or likely to implement Big Data initiatives in the future. In addition, organizations are spending on average $8 million on Big Data in the next year. CEOs, as opposed to line of business leaders, marketing teams, or boards of directors, seem to be the most supportive of Big Data efforts in firms today. Perhaps they see this as a key strategic objective for their firms to remain competitive in the future. Some of the key investments include software, servers, networking, storage, and data and analytics discovery. In terms of where all this data is coming from, the United States ranks at the top in terms of the number of petabytes. This is followed by Europe, Japan, China, and the Middle East. Finally, India and South America are tied at the bottom of the list. In terms of the industries where the most value is derived from Big Data, retail tops the list with both increases in productivity and sales. This is followed by consulting, air transportation, food products, steel, and automobiles. At the bottom of the list, we have industrial instruments, publishing, and telecommunications. The impact of Big Data will continue to grow. It is estimated that the growth in global data per year will be about 40%. This will increase the importance of using analytics to drive business growth and generate operational efficiencies. This increasing importance of Big Data is also evident in the growth in participation from business executives in surveys related to Big Data. For example, New Vantage Partners reported that their 2019 Big Data and AI Executive survey saw a higher participation rate than ever before, suggesting that more executives believe the topic is important. In addition, the survey also documented the following findings:

- 90% of those who completed the survey are "C-level" executives – chief data, analytics, or information officers. A decade ago, only one of these jobs even existed.
- 92% of the respondents are increasing their pace of investment in Big Data and AI.
- 62% have already seen measurable results from their investments in Big Data and AI (a bit less than in 2018, but still pretty good).
- 48% say their organization competes on data and analytics. When Tom Davenport introduced this concept in 2006, perhaps only 5% of large organizations would have said they did so.
- 31% have a "data-driven organization", and 28% have a "data culture". 75% fear disruption from new entrants, and 88% feel greater urgency to invest in Big Data and AI. 92% are driven by positive objectives – transformation, agility, or competition – and only 5% are driven by cost reduction.
- Spending levels are also up in accordance with the respondents' attitudes. The 55% of companies spending over $50M on these resources is up by 15% compared to 2018, and 21% are spending over half a billion dollars on them!
- Virtually all respondents agree that data privacy and cybersecurity are important priorities, and 56% said the same for "data ethics" – probably not even a concept that anyone would have mentioned a decade ago.

The previous points demonstrate the tremendous growth and support businesses have for new Big Data initiatives. Furthermore, a similar survey by TWDI on the state of business analytics revealed that executives were expending benefits from their Big Data investments, ranging from better targeted social influence marketing to manufacturing yield improvements. While there are great growth opportunities with Big Data, there are also several challenges. For example, the TDWI-Predictive Analytics Study uncovered that data integration complexity, initial project selection, data architecture, and hiring qualified staff were some of the greatest challenges executives faced with Big Data initiatives. Lesser challenges included better planning and forecasting, as well as identifying the root causes of cost and manufacturing yield improvements. However, these opportunities and challenges cannot be properly managed if there is a skills gap. A recent survey by the Technical University of Munich identified AI, Blockchain, Novel Interfaces, and the Internet of Things as top skill gaps. To a lesser extent, they also identified Mobile Technologies, Cloud Computing, and Social Media. Laskowski (2015) reported the main factors leading to the failure of Big Data projects in organizations, including strategy failures due to a lack of organizational inertia and/or insufficient use cases. Another challenge is skills failures. These include the Inability to address unanticipated difficulties and the lack of specific and important Big Data skills. Lastly, there are analysis failures, which include not questioning the data properly. On that topic, it's important to ask why the work is important and what the implications are? In addition, there are failures to ask the right questions. These questions could be related to the importance and implications of the work.

These authors believe that one of the greatest promised of Big Data, besides solving some of the most complex and vexing problems in our society, is to provide transparency for all in terms of access to data on government activities. These Data-For-All activities differ from some of the industrial or healthcare applications we discussed, as they are aimed at transparency and providing people with access to vast sources of data. This initiative can actually be considered a modern method of preserving democracy in the West and in any other democratic countries with open data policies. This Data-For-All capability is only possible with the advent of low-cost data storage and processing power, available through cloud computing.

I hope the reader was able to grasp the immense opportunities available to organizations today in using Big Data and AI to help achieve strategic priorities. We live in an age of great change, partially because we have so much data around us and now the technology to make sense of it.

In the next chapter, we will explore how to store Big Data in your organization.

REFERENCES

Fisher, T. 2020. "Terabytes, gigabytes & petabytes: how big are they?" *Lifewire*. https://www.lifewire.com/terabytes-gigabytes-amp-petabytes-how-big-are-they-4125169

Laskowski, N. 2015. "Seven big data failures to watch out for." *Tech Target*.

Rijmenam, M. 2013. "A short history of big data." *Datafloq*. https://datafloq.com/read/big-data-history/239

2 Big Data, Hadoop Distro, Data Triad, and Enterprise Data Lakes

2.1 KEY DATA STORAGE CONCEPTS

- History and evolution of the Data Warehouse Data Marts and Operational Data Stores
- SQL and Big Data
- The evolution and history of Hadoop
- Hadoop Distro comparison: Cloudera vs. Hortonworks vs. MapR
- Overview of Data Triad
- The Role of the Data Triad
- Overview of the enterprise data lakes
- Data lakes vs data warehouses
- Federated model to hydrate the data lake
- Data Ingestion tools and capabilities
- Strategic value of the federated data lake
- Future trends in data storage

2.2 INTRODUCTION

In Chapter 1, we provided an overview of the course and the expectations of the book, the history of Big Data, and reviewed the high-level concepts of where, what, why, and how Big Data and Business converge. We also covered the 5 Vs of Big Data, the promise and challenges of Big Data, and the idea of data for all. In Chapter 2, now that we have defined Big Data, we will now discuss how we store this massive amount of data. In this journey, we will compare and contrast the storage of Big Data with some of the historical ways we have stored data.

2.3 BIG DATA, HADOOP DISTRO, DATA TRIAD, AND ENTERPRISE DATA LAKES

As we embark on exploring the storage requirements for Big Data, it's important to first delve into the history of data used for knowledge along with the technology that supports this referred to as management information systems. Back in the 1960s through the 1980s, information systems in large organizations, including government and private corporations, were primarily used to conduct the regular operations of the firm, which were not largely driven by data and insights. When one of the authors started doing

DOI: 10.1201/9781003190325-2

statistical analysis work at GE Capital back in the early 1990s, in order to get information needed to manage the risk of the credit card portfolio, one had to actually pull data from an IBM 3090 mainframe computer and load it onto either a 3390 Direct Access Storage Device or, if the dataset was large, onto an IBM 3420 Magnetic Tape Drive. This process required time on the IBM 3090 mainframes, which were used to process credit card statements, and computing time was measured in MIPS - an acronym for million instructions per second. In this environment, the analytics did not need to be sophisticated, since even getting a report on the state of the business was information that business leaders were craving. One of the authors once made a simple forecasting model based on ten bands of FICO score, which captured the attention of the CEO of a GE Capital business. Today, this would be considered a simple analysis but back then, it took significant effort just to gather this information. Managing a credit card business required this sort of information in those days. In fact, the Capital One model was an early adoption of business analytics as we know it today. Capital One's origin was Signet Bank, a full-service regional retail bank based in Richmond, Virginia, The founders of Capital One saw that the credit card business, one of the product offerings of the retail bank, could be run more effectively through analytics and did not need the traditional face-to-face relationships that other banking products required. This eventually led to the bankcard division splitting off from Signet to become Capital One. Capital One then became a standalone bankcard business driven by analytics principles. In the early days, before the field of business analytics was established, it was referred to as information-based management. They even went so far as using quantitative reasoning tests for all employees across all functions to ensure that all staff, from human resources to statisticians, could employ quantitative decision-making in their daily jobs.

Eventually, as the world became more digitized, nearly every business began using analytics to make decisions. For example, companies like P&G used to primarily base their analytics on market research studies related to their various famous Brands, including Tide, Crest, Dawn, and Ivory Soap. They shifted from market research to business analytics as they started to have people interact with their web properties and store data from digital interactions in Marketing Data Management Platforms. Similarly, companies across all industries have now begun their analytics journey, especially in manufacturing, with the industrial internet and the Internet of Things. This phenomenon has prompted Microsoft CEO Nadella to claim that "Every Company is Now a Software Company". What he meant was that, in the future, the lines between tech companies and other industries will be blurred. Every company will have a significant digital component, and every company will use analytics as a core component to drive decisions and deliver value to their customers. This transformation will occur over the next two decades, after which companies will need to find new ways to complete. Some futurists predict that the next competitive lever being design thinking, once all companies are on par in terms of analytics. This future scenario contrasts with the way business education is conducted today, which is more quantitatively focused. While we do see some elements of design thinking being introduced to MBA curricula these days, there are still far more courses related to quantitative subjects. Furthermore, many of the design thinking concepts being taught are intertwined with analytics concepts, with the aim of bringing more formality and process discipline to the innovative activities within firms. Firms like

3M have previously balked at combining analytical tools such as Six Sigma into the innovation process, since these tools are often aimed at reducing process variation, which could in turn stifle innovative discoveries. More on this will be discussed in upcoming chapters, but it's still interesting to note in this context.

It's no wonder why businesses must continuously build their capabilities in data and analytics in order to stay ahead of the competition.

The first recorded collection of data in Western Civilization dates back about 7,000 years, when the ancient Mesopotomians recorded crop data on tablets. Fast forward to the industrial Revolution, where data collection first emerged in the textile industry and later expanded to modern manufacturing and government activities. During this period, there was a need for a more systematic and reliable way to store data. Punched weaving looms and eventually punched cards fulfilled this early need.

In the early days of business, post-industrial revolution, entering data into a computer system was a big deal. IBM devised the punch card system to do this initially.

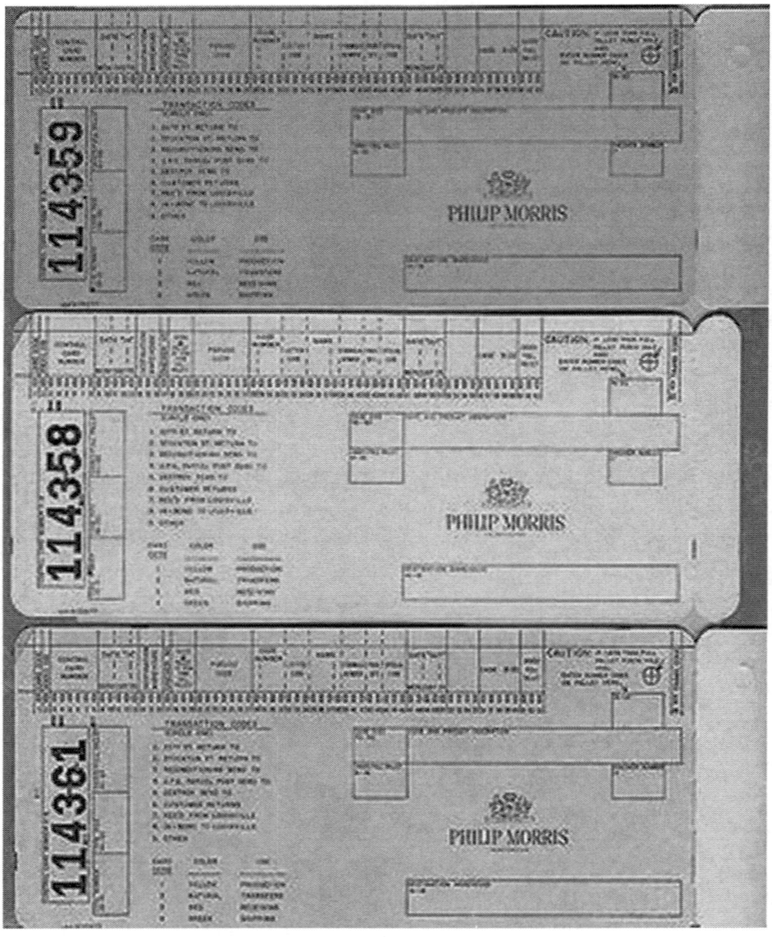

FIGURE 2.1 Early punch cards.

Source: bUS National Archives. Retrieved from: Archives.gov

FIGURE 2.2 Early storage devices - magnetic tape.

Source: US National Archives. Retrieved from: Archives.gov

The origin of punched card systems dates back to the industrial revolution in the late eighteenth and early nineteenth centuries, when they were used to "program" cloth-making machinery and looms (IBM). Later, Herman Hollerith used them with his tabulators in the 1880s and 1890s, which became a core product of what would eventually become IBM. In fact, IBM stands for International Business Machines. Machines are exactly what the early products offered by IBM were. Herman Hollerith's Tabulating Machine Co., based in Washington, DC, developed the tabulating machine to count the 1890 US Census. It sorted and counted information recorded by punching holes in cards (the first punched cards), and Hollerith began selling the machines to retailers, railroad companies, and governments. The method of entering data onto the punched cards was accomplished using punch card machines like the one depicted below. Cards or tape were originally thought as objects to record information, but Hollerith was inspired by how railroad conductors were able to record passenger descriptions using punched cards. For the Census data entry, people were employed to transfer data from paper surveys to the cards. At first, there was no digital storage, so the information from the cards was processed in successive cycles, each involving more complex calculations. After the digital storage revolution, the punched cards started to fade out, but they lasted well until the 1980s. This is a strange phenomenon because even as digital storage and completer entry clearly replaced the need for punched cards, existing business processes continued to operate in a number of firms. A similar situation exists today with the continued use of COBOL language in some firms, despite more intuitive and useful computer languages having replaced it.

As punched cards gave way to binary code, there were sequential storage devices in the form of tapes that would magnetically encode the data. With these tapes, the data was stored and retrieved in sequences, meaning that when accessing and/or storing data on one of those tapes, they had to be spun through the whole tape to find the desired information. The process of reading and writing data on these tapes would be conducted through a language known as JCL or job control language, which was created by IBM. During the magnetic tape storage era, the JCL code would be sent to an operator who would locate the tape that needed to read this file and then load that tape on to the reader machine, and that would load the data into the mainframe computer. Much of the programming on these earlier "powerful" computers was done using COBOL, which is an acronym for Common Business-Oriented Language. COBOL, originally invented by pioneers such as Dr. Grace Hopper, became popular as a compiled language since its instructions could be written in English.

Eventually, magnetic tapes for data storage were replaced by digital storage devices with read/write capabilities. These were very expensive at first, but over time, with companies like EMC, they became lower cost and a much more efficient way of storing data. This gave rise to the era of relational databases and the growth of companies like Oracle, Sybase, and Teradata. With all of this data being stored and processed, there was also the rise in computing power needed to analyze all of the data. Cray Supercomputers were eventually taken over by Sun Microsystems. These fast (at the time) machines relied on the Unix Operating System. The Unix operating system, created in the late 1960s, was popular due to its portability (the ability to run on many different systems), multitasking and multiuser capabilities, and its extensive library of software. It also provided a common command line interface called the shell and provided a common programming interface for the C language. The latter fact enabled most Unix-like systems to run the same application software and desktop environments. These Unix systems, in the early days, were the powerhouses of being able to run queries and process data from relational databases being created from the legacy systems of companies. At first, these systems were add-ons. However, as companies became more reliant on the data and information flow throughout the organization, they became requirements for every corporation. In the mid-1990s, data warehouses were a goal of most corporations, and by the mid-aughts, virtually every large- to medium-sized corporation had a data warehouse stored in a relational database. These databases were typically queried using SQL or structured query language. New IT jobs were created during this era, including the database administrator, who would manage the ETL (Extract, Transform and Load) and the ongoing maintenance of the database.

FIGURE 2.3 Early card punch machines.

Source: Cruz, F. da. (2023). Early Card Punch Machines. Columbia.edu. https://www.columbia.edu/cu/computinghistory/oldpunch.html

The main topic of this chapter so far is how data gets stored. When creating a Big Data strategy, the first thing you need to decide is how to store the data and how to access it. Firstly, we're going to discuss about those topics related to the history of the data warehouse, data marts, operational data stores, and, of course, structured query language, better known as SQL.

During the era when data was coming off of a mainframe, a colleague of mine named Bill Inmon started to discuss the need for a separate system for management information. Prior to this, Dr. Grace Hopper, an earlier computer pioneer (who invented the computer language COBOL and coined the term "debugged") talked about the real benefits of expanding computing power and data storage was to unlock the value of information flowing through organizations.

An influential person who followed Dr. Hopper's advice to focus on the value of information was Bil Inmon, a consultant and considered to be the Father of Modern Data Warehousing. Bill was the first to point out that businesses needed a separate infrastructure and database outside of legacy systems to be able to analyze their data. This paved the way for the development of the modern data warehouse. The author had a chance to work with Bill in the early Ought's, where he was brought in as a consultant to GE to help convince my IT partners that the data warehouse should be less production-oriented and have fewer controls to allow the users to freely conduct their analytics. This was a difficult discussion during this period, as earlier, when Inmon first suggested the data warehouse, the use of information was fast becoming more important for organizations, and IT organizations were trying to figure out the best ways to protect this valuable information for the firm. As this point in time, Information technology groups were "productionizing" the second and third generations of data warehouses, meaning they were treating these entities as something that needed structure and governance. In the early days, the IT organization protected the legacy systems and let the analysts do whatever they wanted with the data extracted from the systems. This gave analysts, in the early days of data science, a lot of freedom in terms of using the data to extract information that would benefit firms. However, the tradeoff was that analysts spent the majority of their time preparing the data for analysis. This author remembers his work doing analytics in the early 1990s at GE Capital, where running data analysis on the mainframe took two days of processing time, including prepping the data in SAS. Overall, this resulted in 80% of an analyst's time being spent on data preparation and 20% actually pertaining to analysis work.

IBM was one of the first companies to create a database specifically for management information. This was known as DBMS, which stands for database management system. Many companies by the early 1990s started to think about moving some of their data off of legacy systems and onto an MIS database platform. When one of the authors worked for GE Capital, it was 1992 when the company first began to allocate resources to build its first data warehouse. Originally, it was looking to outsource its data management to one of the earlier data management software companies that eventually became part of Oracle. This first application was a partnership between marketing, analytics, and IT. This type of partnership, when it comes to developing data warehouses and data lakes, is still relevant to the present time. Whenever there is an imbalance in the decision authority or influence of one of these partners, a problem can occur.

The author got to experience this again when I took an analytics assignment for the General Electric Company in the United Kingdom. The business was a recently acquired private label credit card operation, and when I arrived in Leeds, UK, the mission was to improve the analytics capabilities of the firm. It was soon discovered that we first needed to take the data off of the legacy systems and get it into a form where we could start doing the analysis. In order to do this, they provided me with a budget, staff, and some IT partners (similar to the earlier project at GE). At that time, relational databases were just starting to take hold, so we chose the more expensive route, which was to store the data in SAS, or the Statistical Analysis System. SAS is a proprietary software developed by Jim Goodnight, a statistic professor at the University of North Carolina. Dr. Goodnight and his fellow academic colleagues recognized that there was a lack of statistical software packages with the breadth and scope to handle complex and large studies. Therefore, they decided to develop a comprehensive statistical package known as SAS using an early Bunker Ramo computer. Bunker Ramo manufactured the AN/UYK-3 General Purpose commercial computer in the 1960s and 1970s, which was one of the first commercial computers to use resistors as opposed to vacuum tubes, which has been used previously for commercial computing applications. The application for this software was to more efficiently and effectively analyze agricultural data. The project was operated by a consortium of eight land-grant universities and funded primarily by the United Stated Department of Agriculture.

Dr. Goodnight, along with another faculty member Barr, became project leaders for the development of the early version of SAS. When the software had 100 customers in 1976, Dr. Goodnight and three others from the University left the college and formed the SAS Institute in an office across the street from NC State. Using government grant money and the high-speed computing environment available at North

Carolina State University, they then developed one of the most successful independent software companies on the planet. This SAS software is critical in our Big Data journey since Dr. Goodnight and his colleagues developed not only a 4th Generation statistical programming language to perform statistical analysis but also a proprietary database used to store data. This creation of this proprietary database known as a SAS dataset, was required to be able to use the SAS system. In other words, the data would need to go through this preparation to be turned into a SAS dataset. Once in this format, the power of the SAS programming language and statistical routines could be fully leveraged. The great thing about SAS was that it could extract data from any platform and convert it to SAS, and therefore, in the early days of data mining when there were no data warehouses or any other forms of management information systems, SAS was critical to doing data analytics. Of course, it didn't come without a price both in terms of maintaining SAS datasets and the time it took for analysts to extract the data from a legacy platform, load into a SAS dataset, and then transform the data to be ready for statistical analysis. Some estimates point to greater than 80% of an analyst's time being spent on data preparation as opposed to analysis. The author can attest to this fact, being an analyst myself over this time period. However, much of this effort and time was a result of slow computing times on the mainframe and the consequences of this with errors in the computer code. One of the authors can remember a time where he would have to come into work on weekends to run a SAS job on an IBM mainframe (there was no VPN and high-speed internet and only a very slow dial-in modem). This job would have to be monitored because if there were any coding errors, it could take a few days to run again. Analysts would avoid this by running sample jobs first. However, except for identifying coding errors where the program would not run, if the data was complex, the programmer/analyst would not know whether a condition was not left out of a subsetting "if" statement until the audit after the program was run, so this would take several iterations at great cost and time. These challenges are not the same for today's programmer/analyst, which is a great advantage since more time can now be spent on analytics.

The author can also remember when he first arrived at his international assignment in Hong Kong, where there was a data warehouse in the final stages of development by the IT Group. The problem was that there were no planned users for the data warehouse. Some of the earlier committed parties looking to use it were not even participating in the early rollout phase. As the analytics leader, this author was contacted by the CFO who had agreed to fund the project and was asked to find a way to make use of the new data platform for modeling and analytics purposes. This was not his first time at this rodeo, as he had previously done this both in the United States early in his career and in the United Kingdom on his first International Assignment. Therefore, he became involved in the project and was able to modify its parameters such that he was using the data as a raw material to support all of his analytics. Lucky for him, the project enabled the CFO to effectively get a good return on investment for the project and also served as the author's key data source in order to make an early impact in the international businesses early in his overseas assignment. This also proved to be advantageous for him later in his career when the International CMO actually was later promoted to lead a major division of the company as CEO and then eventually became the Enterprise CFO.

One of the challenges of these data warehouse projects, with development being managed by IT, included the fact that they took longer than the demand for analytics could sustain in the organization. This is where SAS plays a role, as analysts can still access data from legacy platforms while the data warehouse is being built. These types of activities used to take up a great deal of the (then expensive) storage space. IT groups used to look forward to sunsetting these SAS datasets as soon as the data warehouse was completed. SAS got wind of this trend and decided to focus on the database business sometime in the mid Ought's. This was a challenge for them since the SAS proprietary system, despite its links to statistical analysis, was always more expensive. The challenge was that when SAS decided to enter the database realm, their client base converted from statisticians to IT personnel. It's very easy for SAS to convince statisticians to pay a premium for their software since statisticians fully understood the value of their software, and in the early days of data and analytics where the data was distributed among many different legacy platforms, there were no other options for conducting analytics than SAS. SPSS (Statistical Package for the Social Sciences) was available; however, this was not a scalable application at the time and only became scalable after 2009 when IBM purchased the software and made it scalable in conjunction with DB2. Of course, open software statistical computing languages like R and Python changed the entire value proposition for even the statistical computing aspects of both SAS and SPSS later on, but earlier than this, SAS faced a challenge from IT professionals as they expanded into the data storage world. IT professionals are mainly looking for efficient systems to store and retrieve data. SAS solutions are neither cheap nor very efficient, and IT professionals don't have the remit to make the tradeoffs that analysts make when they need to develop solutions even if they are less efficient. When this author worked at the General Electric Company, the run rate for SAS licenses was $20 million, which represented the biggest SAS customer. The culture of the General Electric company was all about improving the bottom line, and when the IT leaders at GE were dealing with SAS, they obviously expected discounts, being their best customers. SAS would not budge on their position of offering a substantial discount, however. This was exacerbated when the Chief Technology Officer of GE invited Jim Goodnight, the founder of the SAS Institute, to GE headquarters, which was located at that time in Fairfield, Connecticut. Despite a face-to-face visit (which only lasted less than an hour), GE was not able to get Goodnight to agree to provide the usual discount GE was expecting off the list price of any software it purchased. What GE failed to realize is that Goodnight, who flew to Connecticut on his own private jet and was the sole owner SAS (which was the largest private software company in the world), did not need to pay heed to GE's aggressive vendor negotiation tactics, which were aimed at convincing vendors that they don't want to lose the GE account at any cost. This could work in some cases. For example, there was a business colleague in healthcare insurance who, early in his career, landed the GE account for one of the largest healthcare insurance plans. When he landed this account, it totally changed his life as the plan promoted him and gave him all of the resources to make the relationship work between the two companies. He also took up residence in Fairfield, CT, which was the town where GE Headquarters was located. GE relied on this type of importance to leverage the best pricing from vendors, but SAS was not interested in any of this.

At one point, the folks in Corporate IT were so frustrated that they aimed to give me and some partners in GE's Corporate R&D $20 million dollars (which was the SAS current run rate). Corporate R&D had some of the world's greatest data scientists and computer scientists, and if anyone could build new statistical software to compete with the SAS Institute (both internal and external), it would be them. This, when combined with my great industry knowledge about SAS and especially how to apply statistics in business, gave us a very good chance of being able to pull this off. When SAS learned that for the first time they had a credible threat, they offered GE an 11% reduction in the run rate. This was the first discount of this type that was ever given by SAS, and for GE, it represented over $2.2 million in cost reduction on at $20 million run rate. Given this discount, GE decided not to invest in creating its own statistical software. I was hoping it would because the project would have been really fun, but unfortunately, this didn't happen. Interestingly, economics would take care of SAS's advantage in this area over the next several decades with open-source statistical computing languages, including Python and R, becoming the most versatile and popular ways to conduct analysis today. Python is built on the C++ base, and R is a derivative of the S language, originally created in Bell Laboratories in response to the fact that, at the time of original development, there were no commercially available statistical packages. So, in true Bell Labs fashion, they created one. This would have been similar to what I did at GE. The S language was first commercialized into S+ in the 1990s and early 2000s, at the time when commercialization of open-source code made sense for the risk that companies would have taken in terms of adopting open-source languages. In the beginning, the original open-source efforts were not businesses at all. They were really groups of revolutionaries who worked together in the fight against the unfair profits that closed-source software companies like SAS, Microsoft, Oracle, and SAP were reaping from customers. Furthermore, these revolutionaries were critical of many of the commercial closed-source software companies' ability to focus on the quality necessary to make their software world-class. They felt that the public was being forced to utilize a substandard product at a fairly hefty price. Therefore, they started with the operating systems and databases, which were representatives of the most broadly used components of software and largely. Through asynchronous collaboration, they authored outstanding pieces of software. This new software was not only open so that everyone could see it, but through a loosely knit governance model, they continuously added, improved, and enhanced it. The first versions of open source were not very user-friendly. This makes sense when one accounts for the fact that the software was originally created by and for developers. However, it was characterized as functional, robust, and flexible. The merits of these open-source creations were eventually seen more widely across the software world to the extent that MySQL started to chip away at Oracle's dominance and Linux became the second most popular operating system for servers (next to Windows). Today, if we look at all the top tech giants - Google, Amazon, IBM, and Microsoft - with the exception of outlier Apple, all of them are either built on top of open source or make use of it extensively. Today, the proprietary software industry is both predator and prey, as open source, once long relegated to the hacker and academic communities, has finally come of age. As such, it is reshaping the software industry as we know it. As a predator, the proprietary software companies are looking to acquire open-source companies. Examples of this include IBM, which acquired Redhat for $32B, Salesforce, which bought Mulesoft for $6.5 billion, Adobe, which acquired Magneto for $1.7B, and Microsoft, which bought GitHub for $7.5 B. These are huge acquisitions, which were unheard of in the early days of open source. However, now large technology companies are seeing acquisitions in this area as protection for their own proprietary software businesses and as huge revenue opportunities. Moreover, these purchases of open-source software by large companies have led them to fully integrate these acquisitions into their strategy. For example, IBM and Redhat wanted to pair up in order to revolutionize the cloud market for business. The idea was to combine Red Hat's open hybrid cloud technologies with IBM's innovation, industry expertise, and sales leadership in more than 175 countries. IBM intends to have this offering based on open-source technologies that allow businesses to securely deploy, run, and manage data and applications on premises, on private clouds, and across multiple public clouds. The business case for this acquisition for IBM is that Digital reinvention is at an inflection point as businesses enter the next stage of their cloud journey. It is estimated that the majority of enterprises today are approximately 20 percent into their transition to the cloud. In this first chapter of their cloud journey, businesses made great strides in reducing costs, boosting productivity, and revitalizing their customer-facing innovation programs. The next chapter, however, is about transitioning mission-critical workloads to the cloud and optimizing everything from banking systems to the supply chain. In order to succeed in the next chapter of the cloud, businesses need to manage their entire IT infrastructure, on and off-premises and across different clouds – private and public – in a way that is consistent, simple, and integrated. Firms require a common environment that they can build once and deploy in any of the appropriate footprints to be quicker to market and more agile. Red Hat has been known for its pioneering business model, which has helped bring open source – including technologies like Linux, Java, Ansible, Kubernetes, Ceph, and many more – to the forefront. Today, it is estimated that Linux is the most used platform for development. Red Hat Enterprise Linux alone is expected to contribute to more than $10 trillion worth of global business revenues in 2022. It is estimated that by 2023, an additional 640,000 employees will be working in Red Hat-related jobs. Some people question IBM's motivation regarding the preservation of Red Hat's independence and neutrality in relation to its open-source heritage. However, IBM has affirmed its commitment to allow Red-Hat to continue its open-source culture, which includes its open-source community leadership, contributions, and development model; product portfolio, services, and go-to-market strategy; robust developer and partner ecosystems; and unique culture. IBM has committed to preserving and protecting Red Hat's mission and unwavering commitment to open source will remain unchanged, and they have also committed to allow Red Hat to continue offering the choice and flexibility inherent in open source and hybrid IT environments. Additionally, IBM has committed to enabling Red Hat to continue building and expanding its partnerships, including those with major cloud providers, such as Amazon Web Services, Microsoft Azure, Google Cloud, and Alibaba.

Other mega-tech companies also have commitments to open-source, including Oracle. Oracle has a long-standing commitment to open source which began in 1998, when the company released the industry's first commercial database on Linux. Then, through its acquisition of Sun Microsystems in 2010, Oracle also became the steward of many other important and long-running open-source projects, such as the Java programming language, which was started in 1996, and the MySQL relational database, introduced in 1995. Today, Oracle is a leader in the development of many of the best known open-source projects in the world, including Java, Linux, MySQL, and VirtualBox. Oracle is also a founding member or Platinum sponsor for key open-source governing bodies such as the Linux Foundation, the Cloud Native Computing Foundation, the Eclipse Foundation, and the Java Community Process. Moreover, Oracle continues to further demonstrate its commitment by running its own internal IT systems for more than 135,000 employees on open source infrastructure, and has publicly acknowledged the benefits of lower IT costs associated with open-source, which is really difficult for a traditional private software company. In addition, Oracle's internal software development organization, which totals more than 45,000 developers, uses many open-source products as part of the product development process. Moreover, Oracle Cloud, one of the world's leading clouds for commercial purposes, is designed with Oracle open-source technologies as a central component. Oracle goes to great lengths to report that, while its support and training departments contribute substantial code and bug fixes to various open-source projects, it is much more than just focused on code support. In addition, Oracle offers world-class support and training for its open-source offerings. Oracle Premier Support provides access to product updates and enhancements, as well as technical assistance and support resources to help firms maintain their products, optimize performance, and effectively implement new software functionality.

This support includes:

- Global, 24/7 access to Oracle experts, knowledge, tools, and best practices
- Continuous access to innovation, including product updates, enhancements, and new releases
- Powerful proactive support resources to prevent problems, speed resolution, and simplify upgrades
- Oracle's unique Lifetime Support Policy.

https://techcrunch.com/2019/01/12/how-open-source-software-took-over-the-world/

IBM was one of the first Megatech companies to commercialize the vision of Bill Inmon and create a database specifically designed for MIS. Advanced analytics were not envisioned at the time, so these databases were developed to facilitate what was referred to as Management Information Reports or better known as reports generated from the systems for management use. It was only later that managers realized that it was not only the facts that could be collected from these systems, represented in reports, but that they also facilitated advanced analyses such as regression, which could be used for such applications like forecasting or modeling for marketing response or analysis and hypothesis testing for marketing and for reducing operational costs. Thus, the Management Information System and its reports became the foundation for what was about to become as analytics, transitioning from being more descriptive to predictive and even prescriptive. No one could have ever imagined in the early days just how far analytics could advance in industry. Certainly, nobody could have imagined the rise of big data and its impact on society at large. Grace Hopper, one of the founding mothers of modern computing, discussed in the 1980s the real power of computers being in the flow of information throughout an organization. At a time when hardware and software were the focus, Grace said, “it’s the information stupid”. This proved to be very true as we moved into the digital age. Interestingly, when the author studied computer science in the 1980s, these texts were literally about hardware and some primitive software. They didn’t even cover data and analytics. Today, as a business analytics professional, the author is called upon to edit and evaluate computer science textbooks to make sure they contain enough analytics content. Even the programming languages today include several high level languages that are devoted to conducting analytics. This was unheard of in the 1980s when SAS was just starting to make its first inroads in this area. SAS and S+ were the only languages in the market for doing statistical analysis, and SAS was just getting started and was not yet in widespread use. Jim Goodnight created the SAS system and programming language out of this need for this type of software to be able to explore and mine data. His original program was academic and used at the University of North Carolina, but he eventually spun it off into a private software company. Given the fact that this software was needed for a growing world of data analysis in businesses, it was adopted primarily by the business world (due to the cost, most academic settings used SPSS). The business world at this time had disparate datasets spread all over the enterprise. SAS was operable on any data platform and created one common data repository. Therefore, SAS became the go-to statistical package to be able to consolidate data into a “SAS dataset” and analyze it. Once the data was in a SAS dataset, many of the best statistical procedures were available for analysts to conduct their analyses. These procedures were developed by statisticians for statisticians, and therefore, proved to be very robust. Most statistical procedures can be found in the SAS/STAT module, with additional procedures under the topic of O/R (Operations Research), ETS (Econometrics and Time Series), DM (SAS Data Miner), SAS Graph, Enterprise Guide, SAS Financial Management, and SAS Insight. Gradually, the influence of SAS waned as more open-source languages, such as R and Python became more prominent. The growth of SAS and the open-source languages R and Python, along with Business Analytics reporting packages like Business Objects, Click-View, and Tableau created a huge demand for analytics and the need to store data and treat it as an asset, just like any other corporate asset. Today, data is not valued on the balance sheet in the same manner as other assets. This deflates the true value of firms who are sitting on a gold mine with their data. Even if the firms themselves are not utilizing the data to their fullest, the process of truly valuing the data may make them an acquisition target or merger partner for other firms that are able to fully capitalize on the data. I suspect we will see much more of this in the future as the value of data takes center stage in corporations. We are already seeing this in M&A activity, including Whole Foods/Amazon, Aetna/CVS, Cigna/Express Scripts, Wal-Mart/Flipkart/Jet, AOL-Time Warner, and Dun and Bradstreet/Netwise/Eyeota. We expect to see much more of this moving forward especially as companies go more digital and depend on analytics to drive their performance. Digital represents the intersection of technology and data, and therefore, the assets of future companies will be in the technical IP and the data.

Interestingly, large companies can borrow techniques from third party data providers who properly value data on their balance sheets. These companies acquire and sell data every day. They also conduct mergers and acquisitions of other data companies, so they have a great handle on the actual value of their data. In a merger, they have to be able to assess the actual value of the acquired firm's value, and if it sells data, then bingo. Interestingly, many companies in the future will sell their data as a service. This is especially true as the Internet of Things fully unfolds and/or more companies sell their software versus or in addition to traditional physical products. These firms, for the first time, will have data to monetize either internally or in the market. It's a good idea for firms today to start the process of valuing their data before it becomes a requirement. That time will come when the data in the firm becomes more valuable than the other assets, including both intellectual property and physical assets. Finance and accounting employees will need special training to handle this emerging asset. Today, the value of the data is either hidden or blended across the book value of other assets. New GAAP standards may also need to be established. The author currently has an advanced data purchase accounting procedure, which deals with the valuation of data and how adding models to the data can increase the value of the data and make it more useful. This brings about some explicit valuations of data and, for the first time, gets the accountants focused on data values. The authors conducted a survey of accountants and asked where the value of the data is captured today. The response was that it was embedded in the other assets of the company, such as intellectual property. The reason for this is that they had no way of knowing how to value the data separately. Some didn't even realize the data could be valued separately from other assets and questioned what the methodology could be. So, therefore, we need to have a training plan for accountants in how to value the data. So far, the techniques for this training are yet to be developed. This book will set the direction and tone to create knowledge in this area, which can then be put forward into practice. This will become an important ongoing discussion topic as we get well underway in the 21st century. Moreover, the author predicts that the proper valuation of data will just serve to increase creating data driven cultures, as more management emphasis will be placed on managing the valuable data assets. These assets will then attract management, who will optimize and make efficient use of the data. Think about how great it would be for a firm to hire a couple of Harvard MBAs to manage the data asset. One firm that did this in the past was ADP, the huge payroll firm. ADT was interested, about 5 years back, in monetizing its extensive data on payroll for companies. For example, healthcare payors would be interested in this data to understand the out-of-pocket costs, as they currently don't receive this information from employers and want to know if the gains they are making in productivity are going directly to the member or not. HR professionals may also want this data in order to benchmark salaries in their industry and be able to slice and dice this information for greater insights. Moreover, HR professionals may be able to benchmark the additional benefits given to employees from competitive firms, as these benefits would most likely be listed on the payroll documents. ADT can add value to the data by properly coding these entries in the data set, making it easier for the end users of the data. This is a great example of a company that previously never thought of the data that is used for their payroll processing as valuable. Some companies like Experian or Equifax have been selling their data from the beginning. However, others, like ADP, are just beginning to monetize their data. Additional companies sitting on a mountain of valuable data include pharmacy benefits management companies which have all the drugs that patients are taking and, therefore, their clinical conditions. What a wealth of information for pharmaceutical firms looking to set up clinical trials and need to find people for testing who have certain health conditions. These firms can then send digital ads to these people via cookies and try to recruit them for clinical trials. The drug company will never take possession of the names of the individuals until they respond to the offer to enter the trial, thus preserving confidentiality. The targeting is not 100%, but it's definitely better than targeting at random. This is especially true for very hard-to-find chronic conditions that are sought after for testing by pharmaceutical firms, where trying to find these people using normal advertising methods would be cost prohibitive, as finding these people is like finding a needle in a haystack. Conditions like ulcerative colitis have only 10 cases per 100,000 people in the population. Therefore, drug companies what to know exactly who the consumers are who have this illness for marketing purposes and/or clinical trials and other research. Companies like DeepIntent have huge databases of these folks. Pharmacy benefits management companies like Optum (United Health), CVS/Aetna, and Express Scripts also have a great deal of information about people who take medications and, from this information, can identify what chronic conditions people have and sell the data to pharmaceutical companies looking to target patients for clinical trials. Of course, specific names and addresses are not given, but these companies can go through a third party like Liveramp or Neustar to match the record with a cookie or a device ID. Once this onboarding process has been completed, the firm using the data can then send out digital media such as display advertising, to attract the right people to the right clinical trials. If the creative is worded in a clever manner, the customers will not even be aware that they are being targeted for the trials since their PII is not present in the ad, but they know. The author, for example, has a chronic GI condition and knows when he is being tracked for advertising the latest treatments, as they begin to show up in an omni-channel fashion as soon as the author logs into his PC. As an alternative, the advertising companies can get depersonalized data with population health data appended and can create models of those most likely to have the health condition. This is also another effective way of targeting. The accountants should be pleased with this as it allows them to better understand the cost of acquiring third party data and what its real value is. Once this is done, it should serve to increase the value of the firm, as the value of the data will be extracted from other assets of the firm and valued separately. So, how will data be valued? Firstly, one would have to examine the current and future use cases. One obvious way to value new data to be acquired is to just look at the data companies like Dun and Bradstreet and Equifax. The companies are buying up other data companies all the time, and each time they make an acquisition, they will be using various methods to value the company to be acquired. These methods can be replicated for internal data sources in a firm. However, as with all valuation methods, there are flaws in that the method depends on the future cash flows from selling data, which firms wanting to value their internal data typically do not do. Another way to value new data sources is to conduct experiments by adding the new data to existing analytics applications like models. Then, the incremental improvement from adding the new data can then be valued, and then the firm will know the exact incremental value of the data customized for this specific organization. Of course, the flaw with this one is that it doesn't measure the future options of what can be done with the data. Sometimes, data is not valuable immediately but then can be very valuable when there is a shift in the economy. For example, when credit cards started offering 0% interest rates for a set period of time, new variables that were previously not so valuable in rate-surfing models now became valuable in rate-surfing models. These variables included the number of trades recently opened in the last 6 months. This variable was correlated with rate surfing behavior. Similarly, during the COVID-19 pandemic, variables indicating the use of virtual care became significant due to the shift where people actually engaged in more virtual care. Companies like ADT are looking to monetize their data by selling to third parties. Perhaps this is a good way to fund a data lake consisting of a large amount of data elements, some of which are not currently being utilized. There is also the idea that data alone may not be useful, but when matched with other data (especially external data), it then becomes very valuable. An example of this is data from a company like ADT, which has all of the benefits information on employees of its customers but has no use for this information in its own business. It does not actually know what people are doing with their benefits. Now take a company like Aetna, which has the claims of its customers but has no idea actually how much the customer is actually paying out of pocket for their healthcare benefits. Put the two databases together and you have something tremendously valuable for Aetna. This matchup makes the claims data come alive. Once the data is properly valued, firms can then make the appropriate investments in storage, processing, security, and acquiring new data in order to optimize their investments. A focus on valuation will solve a number of challenges in business, including how to properly purchase and negotiate third-party data. The third-party data industry is horrific for customers looking to get maximum value from their data. For one, they convince purchasing staff to negotiate 3-year contracts, which are troublesome since sometimes the customers find that the data doesn't have much value for them when they put it to full use, and there is no way out of the contract. Furthermore, unlike any other asset that one leases or buys, there is no salvage value, which allows one to recover some of the money spent on the asset. The third-party data would not even let the business put the use of the data up for auction, since this would be competing with their own sales force. What the authors recommend is that the data not be purchased in advance but to test it first and then seek to find the incremental value. Then, go back to the third-party data provider and show them the true incremental value of the data. You should refuse to purchase the data if the price exceeds the incremental value of the data, since this action would be detrimental to the shareholder. The sales force of the third-party data company will be resistant to this action, since the commission on data are some of the highest in industry. The commission on third-party data is well north of 10%. With the average data contracts of some of the top firms being greater than $2 million per annum, this means that these sales folks can earn more than $200,000 per year from just a single client. Many of them have 5 or more clients. Therefore, you can see that they have an incentive not to lose any margin on their product. You need to establish this process upfront before you make the initial purchase. Otherwise, it will be very difficult to do once the commissions are flowing into the bank accounts of these salespeople. These sales folks, once successful, set goals on what they are going to purchase with their commissions in advance, and the parent company already books the income, so there is no going back once you sign contracts. They will also sue you if you refuse to pay. This could cost more in legal fees than it's worth.

One other consideration that really makes these sales personnel happy is making an advance payment or prepayment of the data. Please note that it's technically a prepayment but not exactly. It's actually the conversion of data (which needs to be amortized over the future year and cannot legally be prepaid under GAAP). Of course, all of this is predicated on having enough money left by the end of the fiscal year. The idea is to have the third-party data provider convert the data into a professional service. They do this by first setting the value of the data to zero. Now your third-party provider will be uncomfortable with this because they are staunch defenders of the value of their data. However, the data needs to be set to zero in order for the accounting treatment to work. Then, you instruct the third-party to develop a model to be used to create enhanced data. The model can be anything that adds value to the data. For example, a B2B model could be a decision committee score for B2B purchasing decisions. Or for B2C marketing, it could be a marketing model for acquisition. Or it could be a lifetime value model for a subscription business. Whatever the case, it must be a credible model that the business intends to use. The model development can be billed as a professional service for the value it is adding to the data. More on how this works is if the data will be billed for a million dollars. The value now gets set to zero, a model is built, and the value of the model equals 80% of the original price of the data. The accounts only allow 80% of the process to be billed as a professional service. The model can now be paid for in the current year and used on the "free" data in the next year. Then, moving forward, this can occur every year, and in the year that the funds are to be used, the payment to the data firm can be made in January of the following year. It's a good thing to not let too many funds accumulate in this manner. The author knows of one firm that caps this at around $2 million. The reason for this is that investors would want the additional capital invested or else returned as a dividend. Microsoft once had a large amount of retained earnings invested in banks at a low rate of return. The resulting cash pile, which grew to nearly $60 billion, led to investors pressuring the company to return funds to shareholders; and in July 2004, Microsoft announced the one-time dividend, along with a doubling of its annual dividend and a four-year share buyback plan. While a data prepayment isn't going to convert into the war chest that Microsoft has, nevertheless, Finance professional will get nervous if the sum gets too high, and therefore, it's a good idea to keep it under $2 million, which, for a larger Fortune 500 Company, is a drop in the bucket in terms of budget. This will not attract too much attention from the finance community. However, $5–10 million may be considered excessive, using the Microsoft example. It's really a partnership and a tradeoff at the same time, since in the author's experience, sometimes finance will ask to use the funds to plug a gap, and if the funds are large enough, finance can take a portion and leave the rest for the analytics or data manager. If this is the case or is to be the case, then finance will allow a larger escrow in the first place.

In terms of data storage, unstructured data, which largely comes from web platforms, including streaming entertainment, social media, and online shopping, presents a challenge when using SGL. Therefore, other solutions have been developed to overcome the challenges of using SQL for storing unstructured data.

We will now talk about Hadoop. Hadoop was invented in 2002, when big data started to emerge, and Apache software created a new type of database to handle large unstructured data, which was called Hadoop. The name "Hadoop" comes from an Indian cartoon character, an elephant. The developer of Hadoop had a son who watched this cartoon, which is where the inspiration for the name came from. Unlike structured databases, Hadoop can handle and process large amounts of unstructured data, making it ideal for the data storage and processing requirements of the digital age. Many firms are now uploading data to the cloud but will keep their most sensitive data on premises. This is where Hadoop comes into play. With on premise infrastructure is not as powerful as cloud technology, Hadoop can process large amounts of unstructured data more efficiently and effectively. With such large amounts of data being processed, firms must take care to make it easier for analysts to understand which data is available.

Experfy.com (2025) points out that, because of the flexibility of Hadoop, it enables the addition of new functionalities for the management of Big Data and the accomplishment of data management tasks. A number of different software companies have taken advantage of this and have made several improvements/adaptations to its functionality.

The three most popular companies are: Cloudera, MapR, and Hortonworks. Cloudera has been around for the longest, and along with Hortonworks, it remains open-sources, while MapR contains proprietary components. Each of these vendors have overlapping features, along with their individual strengths and weaknesses. Cloudera has been here for the longest since the creation of Hadoop. Hortonworks came later. While Cloudera and Hortonworks are 100% open source, most versions of MapR come with proprietary modules. Each vendor/distribution has its unique strengths and weaknesses, and they all have certain overlapping features as well. If you are looking to make the most of Hadoop's immense data processing power, it makes sense to make a comparative study in the top three Hadoop distributions. Cloudera provides a proprietary management suite to enhance use convenience, including real-time nodes and a reduction in deployment time. Hortonworks, founded in 2011, has many recent innovations, including Yarn, which performs better than MapReduce by enabling the inclusion of more data processing frameworks. MapR's proprietary components provide customers with improved system reliability (detecting and fixing bugs etc.) and provide technical assistance and other expanding functionalities. Below is a table from Experfy.com (n.d.), which compares and contrasts the three:

	Hortonworks	Cloudera	MapR
Performance and Scalability			
Data Ingest	Batch	Batch	Batch and streaming writes
Metadata Architecture	Centralized	Centralized	Distributed
HBase Performance	Latency spikes	Latency spikes	Consistent low latency
NoSQL Applications	Mainly batch applications	Mainly batch applications	Batch and online/real-time applications
Dependability			
High Availability	Single failure recovery	Single failure recovery	Self healing across multiple failures
MapReduce HA	Restart jobs	Restart jobs	Continuous without restart
Upgrading	Planned downtime	Rolling upgrades	Rolling upgrades
Replication	Data	Data	Data + metadata
Snapshots	Consistent only for closed files	Consistent only for closed files	Point-in-time consistency for all files and tables
Disaster Recovery	No	File copy scheduling (BDR)	Mirroring
Manageability			
Management Tools	Ambari	Cloudera Manager	MapR Control System
Volume Support	No	No	Yes
Heat map, Alarms, Alerts	Yes	Yes	Yes
Integration with REST API	Yes	Yes	Yes
Data and Job Placement Control	No	No	Yes
Data Access			
File System Access	HDFS, read-only NFS	HDFS, read-only NFS	HDFS, read/write NFS (POSIX)
File I/O	Append only	Append only	Read/write
Security: ACLs	Yes	Yes	Yes
Wire-level Authentication	Kerberos	Kerberos	Kerberos, Native

FIGURE 2.4 Table comparing Hadoop offerings.

Source: Experfy.com (n.d.) Retrieved from: https://resources.experfy.com/bigdata-cloud/cloudera-vs-hortonworks-comparing-hadoop-distributions/

Next up in this chapter is a discussion on the Data Triad. The data triad, better known as the CIA data triad, stands for confidentiality, integrity, and availability. All data storage principles discussed in this chapter should follow this model, or something similar to prevent critical and/or sensitive information from being breached. The CIA data triad is an important concept in the information security industry. The

GDPR mentions it in article 32, and it is used in ISO 27001, a global standard for managing information security (IT Governance, n.d.). Confidentiality, the first element of the CIA triad, relates to keeping sensitive information secure and private, and deterring cyber criminals or employees from accessing data without authorization. Integrity, the second concept in the CIA data triad, refers to the firm's ability to protect the data from corruption, namely the completeness and accuracy of the data. Most people think data protection as being related to preventing unauthorized access, but equal importance should be placed on protecting the integrity of the data. The final and third element of the CIA data triad is availability. Data availability refers to a firm's ability to access information when it is explicitly needed (IT Governance, n.d.).

There are specific databases for unstructured data. One particular is the NoSQL database. As the name suggests, NSQL database do not depend on structured data and do not use SQL to enable access. If one desired a similar setup to a structured database, tools like Grok can be used. Grok is a filter tool within Logstash that is used to parse unstructured data into something structured and query able. It is used extensively for working with log files.

Now let's discuss the Data Lake. Bill Inmon was the first to conceptualize the idea of a data warehouse being a system designed solely for decision support. The Data Warehouse was an enterprise tool, so Data Marts were created to conduct analytics for specific functions or applications. The Operational Data Store was also created to house raw data in process, which would then flow into the Data Warehouse and other applications. However, with the rise of Big Data, the concept of the data warehouse became somewhat outdated. This is where data lakes come in. Data lakes can handle the massive unstructured data associated with the digital age and big data. They can be housed on premises, in the cloud, and also in a hybrid setting. One of the challenges of the data lake is the creation of a data landfill. Data landfills occur when data accumulates in a data lake, but users never access a large portion of it, leading to unnecessary storage costs. The challenge for firms is that they may not be able to predict the future use of the data, which makes them reluctant to delete anything. This is a dilemma not easily solved. Luckily, cloud computing has significantly lowered the cost of data storage, enabling firms to retain large amounts of data even if they have no short-term intentions of using it. However, firms still need to balance the amount of data they store and with its future or current use, as more data means more work for analyst sorting through the jungle to mine useful information.

Now let's talk about how we ingest data into the data lakes. One approach is the federated model of data ingestion. Domo (n.d.), reports that the federated data model keeps the data in its place of origin while using virtualization techniques to unify the data view and access to information. This is in place of other models that consolidate information or remove sources. In essence, data federation makes the data easily accessible and creates a virtual single data source. Some people tend to think of data lakes as having all the data consolidated into one single source. In fact, the term "Data Lake" can mislead people into thinking of it as a large lake (like Lake Superior or Lake Michigan) filled with data. Some people are under the false impression that data consolidation and data federation mean the same thing. This assumption would be wrong, as data consolidation involves converting data into a single source that is then stored in a separate, fixed location (i.e., in a data lake or a data warehouse). The question then becomes: Can a data lake be federated? The answer is yes, of course. A data lake. A federated data lake is a scaled interconnected pool of data where users can access through single queries. Science Logic (n.d.) reports that: "Data federation, plus the functionality of a data lake, not only allows for the analysis that delivers insights into the state of the infrastructure and applications that support the business, but also enables remediation of issues affecting performance and automates IT processes and workflows in real time".

Now we will talk about the strategic value of the federated data lake. Federated data lakes are a key data management strategy in the world of Big Data. This is because, with the high velocity of unstructured and structured data being collected in real-time, it's critical to have the data accessible at the right time. Federated data lakes can be used across all major industries to improve customer interactions, as customers in the digital age have come to expect firms to provide critical information that aligns with their specific preferences or interests. The federated data lake can help firms deliver the raw materials to meet the customer's expectations in the most efficient and effective way possible.

2.4 CONCLUSION

Now, we will end this chapter on data storage with a brief look at the future trends in data storage. Firstly, we have taken the reader through the storage of data from the first recorded collection of data in Western Civilization was about 7,000 years ago, when the ancient Mesopotamians recorded crop data on tablets, to modern federated data lakes stored on the external cloud, on-premises, or the hybrid cloud. What's next? Firstly, we have AI integration, where AI will automate data management tasks like data protection, provisioning, and capacity planning. The next trend involves higher-capacity storage devices like solid state drives or higher capacity tape drives. Moreover, DNA storage is expected to grow significantly. Storing data on DNA molecules is an emerging technology trend that greatly improves the efficiency of data storage. There are new developments in writing technologies, such as Shingled Magnetic Recording, Perpendicular Magnetic Recording, Enhanced Caching, and even installing helium inside the casing will enable the capacity of HDDs to rise to 100TB by 2025. Finally, new generations of NVMe storage are coming out. NVMe is a recent storage access and transport protocol for flash and next-generation solid-state drives (SSDs). NVME delivers the highest throughput and fastest response times for all types of enterprise workloads.

REFERENCES

Domo.com. n.d. Retrieved from: https://www.domo.com/learn/article/a-guide-to-data-federation-everything-you-need-to-know

Experfy.com. 2005. Retrieved from: https://resources.experfy.com/bigdata-cloud/cloudera-vs-hortonworks-comparing-hadoop-distributions/

IT Governance. n.d. Retrieved from: https://sciencelogic.com/blog/data-federation-and-the-modern-enterprise

Science Logic. n.d. Retrieved from: https://sciencelogic.com/blog/data-federation-and-the-modern-enterprise.

US National Archives. Retrieved from: Archives.gov

3 The Data Supply Chain

3.1 INTRODUCTION

Traditional manufacturing firms like Ford Motor Company used to manage complex global supply chains to ensure that they had the raw materials needed to develop their final products, which in this case were automobiles. This supply chain continued right up to the assembly lines. Ransom E. Olds of Oldsmobile fame was the first to create the stationary assembly line, where the automobile came to the worker, as opposed to vice versa. Henry Ford then followed this model. All of the parts on this assemble line came from the Oldsmobile global supply chain. In fact, Olds was the first company to utilize third party suppliers (Florea, 2023). Olds was also another pioneer in interchangeable parts making it the first automobile company to mass manufacture cars with interchangeable parts. This was then followed by Henry Ford, although most people think it's Ford who was the first to use interchangeable parts. Oldsmobile was eventually purchased by General Motors and became one of its operating companies but still remained innovative. Later, it was the first vehicle to utilize an automatic transmission. With the Rocket V8 engine, Olds also produced one of the fastest cars in the post-WWII era, especially in the production of its legendary 4–4–2 muscle car. Finally, by the 1960s it had produced the first turbocharged production car (Florea 2023). Henry Ford took his supply chain seriously. One example is that from 1928 to 1945, the Ford Motor Company in an effort to ensure a stable supply of rubber for tires and other parts critical to the manufacturing of automobiles, built the Fordlandia and Belterra plantations in Brazil (The Henry Ford, 2015). Prior to the Fordlandia and Belterra plantations in Brazil, the rubber industry was an oligopoly between the Netherlands and England due to their colonies in South Asia. This kept the prices for rubber high in the market. Interestingly, rubber trees are native to the Amazon rainforest, but the British transferred seedlings to the East Indies and used this to corner the market (The Henry Ford, 2015). By 1941, there was an estimated 3,651,500 rubber trees planted at Belterra and Fordlandia (The Henry Ford, 2015). At this time, nearly three-fourths of the rubber imported into the United States was used to manufacture automobiles. Therefore, the US government and the private sector, in particular the automobile manufacturers, started looking for new rubber plantation locations in order to address the monopolistic prices set by the British (The Henry Ford, 2015).

The need for natural rubber was eliminated with the onset of synthetic rubber, which received accelerated development after WWII when Japan seized the rubber plantations in Southeast Asia (Nyla, 2022). Not only was rubber a big supply component in the manufacture of automobiles as described above, but it was also critical for war equipment (i.e., jeeps, gas masks, bombers, and tanks) and critical to winning the war. The Germans were the first to create synthetic rubber. This created a big threat for the US, as if there two enemies in WWII were the only ones with access to rubber needed for a mechanized military. This resulted in the US stepping up efforts to develop their own version of synthetic rubber. Nyla (2022) discussed that B.F. Goodrich built a

DOI: 10.1201/9781003190325-3

trial facility for generating butadiene-copolymer synthetic rubber in 1937. It employed synthetic rubber in several of its commercial goods within two years. However, synthetic rubber remained a costly option as long as natural rubber sources were cheap and plentiful. It is interesting to note that the Germans also found synthetic rubber to be expensive to produce. In June 1940, B.F. Goodrich developed the first synthetic rubber passenger automobile tire in the United States. This represented a significant win for the US, as it was now on par with German synthetic rubber manufacturing. The B.F. Goodrich Company went on to become the leader in the rubber sector and its related products throughout the first half of the 20th century. For example, vinyl was also invented by a B.F. Goodrich scientist. In the town where the author grew up, in Shelton, CT, there was a huge B.F. Goodrich plant that produced sponge rubber mattresses. This was a huge plant that brought thousands of jobs to the town. In fact, the entire region, known as the Naugatuck and Housatonic Valley, developed a competency around rubber due to the fact that Charles Goodyear, a local resident, was the inventor of vulcanized rubber. Goodyear obtained a US patent on the vulcanization process in 1844. Vulcanizing rubber involved chemically treating it to prevent the rubber from becoming sticky in the heat and brittle in the cold (Fogarty, 2014). In addition to rubber mattresses, the plants also produced Keds sneakers. In fact, when the author was growing up, in a time before stricter EPA regulations, one could tell the color of the sneakers being produced that day by looking at the color of the Naugatuck River as the water fell over the various dams, as the factories dumped their waste directly into the local rivers (Fogarty, 2014). There was a great deal of innovation going on in this area around rubber products. For example, the artificial leather product called Naugahyde was developed by US Rubber and named for the town it was located in, Naugatuck, CT. Naugahyde was the first rubber-based artificial leather ever made. Naugahyde is currently a subsidiary of Uniroyal Global, which was previously known as US Rubber. It is now recognized as a global leader in high performance vinyl coated fabrics and topcoats. The above paragraph shows how complex supply chains can be for just a single product that goes into auto manufacturing, namely rubber. While the US still has a strong manufacturing base, including automobiles, there is also a faster growing service economy. Service economies don't typically have a complex supply chain like the one described above with automobiles. However, one of the key aspects of today's successful service business is the use of data to generate insights, which can help the service business be more efficient and effective. This gathering of the data in order to produce insights in modern organizations can be thought of as the data supply chain, and in many cases, it can be just as complex as a manufacturing supply chain. I outlined the rubber supply chain for automobiles above as a call to action for data managers to treat their data supply chain with the same importance that Henry Ford did with rubber. While many people would not assume that a data supply chain would not lead to an important outcome similar to the creation of an automobile, this author begs to differ. One example of an experience the author had was in relation to financial services when he worked for GE Capital. At the time, the owner started patenting some of the models that were developed to improve the marketing and risk management of products. This was directly in line with one of the goals of the parent company, General Electric, at the time, which was outing quotas on patents for each business as it wanted to follow the IBM models, which made billions from licensing

US007305364B2

(12) **United States Patent**
Nabe et al.

(10) **Patent No.: US 7,305,364 B2**
(45) **Date of Patent: Dec. 4, 2007**

(54) **METHODS AND SYSTEMS FOR SUPPLYING CUSTOMER LEADS TO DEALERS**

(75) Inventors: **Oumar Nabe**, New York, NY (US); **David J. Fogarty**, Riverside, CT (US); **Ian R. Bell**, Peebles (GB); **Balwinder S. Samra**, Yorkshire (GB)

(73) Assignee: **General Electric Capital Corporation**, Stamford, CT (US)

(*) Notice: Subject to any disclaimer, the term of this patent is extended or adjusted under 35 U.S.C. 154(b) by 929 days.

(21) Appl. No.: **09/828,530**

(22) Filed: **Apr. 6, 2001**

(65) **Prior Publication Data**

US 2002/0194050 A1 Dec. 19, 2002

(51) **Int. Cl.**
G06Q 40/00 (2006.01)

(52) **U.S. Cl.** **705/37**; 705/10; 705/38

(58) **Field of Classification Search** 700/214; 707/102; 705/7, 10, 37, 38
See application file for complete search history.

(56) **References Cited**

U.S. PATENT DOCUMENTS

5,822,410 A * 10/1998 McCausland et al. .. 379/114.01
5,835,896 A 11/1998 Fisher et al.
5,930,764 A 7/1999 Melchione et al.
5,956,693 A 9/1999 Geerlings
5,963,910 A 10/1999 Ulwick
5,966,695 A 10/1999 Melchione et al.
6,012,045 A 1/2000 Barzilai et al.
6,029,068 A 2/2000 Takahashi et al.
6,049,599 A * 4/2000 McCausland et al. 379/111
6,049,782 A 4/2000 Gottesman et al.
6,061,506 A 5/2000 Wollaston et al.
6,073,112 A 6/2000 Geerlings
6,078,892 A * 6/2000 Anderson et al. 705/10
6,115,691 A 9/2000 Ulwick
6,182,054 B1 1/2001 Dickinson et al.
6,183,362 B1 2/2001 Boushy
6,226,623 B1 5/2001 Schein et al.
6,236,975 B1 * 5/2001 Boe et al. 705/7
6,243,691 B1 6/2001 Fisher et al.
6,321,206 B1 * 11/2001 Honarvar 705/7
6,338,066 B1 * 1/2002 Martin et al. 707/102
6,430,539 B1 * 8/2002 Lazarus et al. 705/10
6,539,392 B1 * 3/2003 Rebane 707/101
6,567,814 B1 * 5/2003 Bankier et al. 707/101
6,658,391 B1 * 12/2003 Williams et al. 705/10

(Continued)

OTHER PUBLICATIONS

Chou, Paul B.; Grossman, Edna; Gunopulos, Dimitrios; Kamesam, Pasumarti. "Identifying Prospective Customers", 2000. ACM Press. pp. 447-456.*

(Continued)

Primary Examiner—Hani M. Kazimi
Assistant Examiner—Peter Choi
(74) *Attorney, Agent, or Firm*—Armstrong Teasdale LLP

(57) **ABSTRACT**

A method is disclosed for generating customer leads for use by dealers attempting to sell a product, for example a loan product. The method is implemented on a system and includes providing a database of customer information, predicting a propensity for one or more customers within the database to respond to an offer, predicting when the customers will respond to the offer, generating a potential customer list, and providing the potential customer list to one or more dealers.

37 Claims, 17 Drawing Sheets

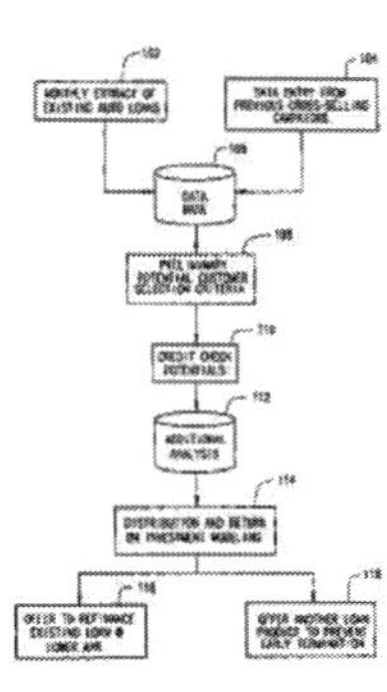

Source: https://patents.justia.com/patent/7305364

its patents. I became a true believer in patenting analytics innovations when I worked for GE Capital, and we sold a financial services business but kept the analytics patents and licensed their use back to the acquiring firm after the disposition. These patents created a continuous income stream for GE Capital long after the business was sold. Below is an example of one of my innovations that was licensed after a disposition:

This demonstrates the true value creation in financial services businesses, which are analytics and insights that competitors do not have, along with the brand assets as opposed to physical assets in manufacturing firms. I will posit that a model developed for a financial services firm is just as valuable as a new car model coming off the assembly line at an auto manufacturer. This author has experienced models in a past life, which have enabled the firm to generate over half a billion in annual profits. The models were developed and implemented in place of human underwriters for consumer loans. The models were better able to optimally get the weights for the variables related to loan default and, therefore, resulted in a superior risk assessment than humans and, therefore, enabled incremental loan business to take place. In healthcare, these models can help to identify high-risk patients to recommend certain preventative treatments, thus saving providers or other healthcare entities hundreds of millions of dollars. Contrast that with the development of the new Ford. For example, Ford sells about 127,000 Broncos per year. At an average of 3,000 per Bronco profit, that's nearly $400 million in profits for Ford's Bronco line. When this author worked for GE Capital, the company acquired a Japanese personal loan business, and models were used for the first time ever at this company to replace their manual underwriting. The models enabled the increase of credit lines across the portfolio for lower risk-of-default customers and decreased the credit lines for higher risk of default customers. This activity and the use of the models eventually resulted in over $500 million in incremental profits for the firm. So, as it can be seen from this comparison example, the use of analytics and models in a service business can be just as valuable, from a profit perspective, as introducing a new product in a manufacturing business. Therefore, the treatment of data as a supply chain to ensure a steady supply of the raw material in order to produce the analytics product is fully justified.

Davenport, Evgeniou, and Redman (2021) highlight that supply chain management techniques may be a partial solution to the challenges that many organizations face with respect to the management of their data. In this article, the authors discuss that companies have always produced data products in terms of spreadsheets, reports, etc. However, now the products are getting much more sophisticated, examples of which now include decision engines and chatbots. They then outlined the key steps to establishing a data supply chain. The first is for the Chief Data Officer in the firm to appoint a Data Supply Chain Manager from their staff. The second step is to document the data and associated cost, time, and quality needed to deliver the various data products the firm produces. The next phase is to define the data supply chain, develop high level and detailed flow charts, and establish flow metrics. The third phase is then to establish process metrics. Metrics like process inventory, flow rate, and flow time (or throughput), along with process capacity and utilization, are great places to start. Once we have these metrics then we can improve the process, such as identifying and eliminating bottlenecks. Bottlenecks represent the part of the process with the lowest capacity. Since it's the lowest capacity, identifying it and finding ways to expand capacity can improve the throughput of the process. Some bottlenecks for data supply chains that the author has experienced include data masking required to send sensitive data and performing data quality initiatives, which can include data transformations and imputations. Both of these activities add to the time for an analyst to be able to convert data into insights, which could then be utilized by the business. The key to any insights organization, especially one undergoing a digital transformation, is the rapid generation and consumption of insights. These insights will both fuel the pace of the digital transformation and create a competitive advantage for the firm using the digitally collected information for the first time. The author had a situation where, due to lack of data quality, there was a bottleneck for auditing and signing off on the data. This resulted in up to a 2-week delay in obtaining the insights, much to the dissatisfaction of upper management. A two-week delay in providing insights, which can be used to improve the product or service offering, is huge in the digital age. This was even more true in the author's industry, which was Medicare insurance, as the national enrollment period was only a few months for most individuals. There were also cases where insights that were produced had to be restated since quality errors were found after they were released. This made management even unhappier, as they had already reported the numbers up to the Executive Leadership and restating numbers that had already been booked and used to decide things their executive compensation is not good. Some organizations also present these numbers to Wall Street analysts eager to see progress in the digital transformation of the firm. In this case especially having to restate becomes a credibility issue. So, therefore, these cases demonstrate that delays in the timing of insights can have a significant effect on a firm.

Given the above paragraph, we can think of the creation of insights as a process and can view the creation of a useful insight for the business as the process flow metric. Our goal is to improve the quantity and quality of insights for the business. Please note that some analysts may reject this approach, saying that you cannot really increase the quantity of insights since you don't know whether you are going to get true insights at all. They will comment that creating insights is not like baking a cake. While this is true to some extent it is also a fact that insights can be turned into a process similar to creating a process for innovation. Of course, discoveries like the Post-it Note or Velcro may not happen, but companies like Apple have proven they can create a process to have a steady stream of innovations even after the passing of their founder. Microsoft is another firm that no longer has breakthrough innovations like the introduction of Windows, but instead now has a steady stream of innovations improving upon the original breakthrough innovation. Actually, the Microsoft example is an interesting one. For example, when Microsoft launched Vista, it was considered a breakthrough. However, none of the customers really liked it, from both a feature and performance perspective, so Microsoft went back to an enhancement of the original platform with Windows 8. This goes to the old saying that, in some cases, you can't build a better mousetrap.

The author understands the tradeoff between the art and science in the creation of insights. The discipline of having to create insights under a timeline can actually help train analysts to produce better insights. Some analysts tend to want to take the easy way out and prepare reports without extracting true insights. They expect the business leaders to be able to look at the report and come up with the insights. When this is the attitude, there is no risk put on the analyst. You see, when you interpret the results of an analytics exercise, even though it is based on solid numbers, one is still expressing an opinion on how to interpret the data, and this represents a risk in terms of one's professional abilities and even their intelligence, which would prompt them to question their own abilities as analysts if they missed something. This is especially the case if the manager is able to identify it, as the analyst could be embarrassed that his/her manager discovered the insights when they might feel it was their job to do this. However, it is the long-term experience of this author that everybody comes to the data with a different perspective, and therefore, just because your boss, without the PhD, has good insights should be considered a plus rather than a negative. I always try to get as many perspectives on the meaning of the data as possible and consider my role to present and validate those different perspectives. My analyses have become much more powerful over the years, having these great and varied perspectives. On top of that, the non-analytic managers contributing to the insights feel that they have been part of the process and are more likely to adopt the implications for implementing the insight. For example, this author had a very senior manager in the management line, several levels up, who would always like to see the raw data and extract her own insights. She had a very powerful sense of what was happening in the data and used this to establish a direction and strategy for the organization. The author must admit that sometimes he didn't even catch the insights that the manager derived, even though he had decades of experience in extracting insights from the data. This is a good process that everybody responsible for producing insights should take on as a best practice. Insights are everyone's responsibility in modern organizations, especially when there is a variation in business and marketplace understanding among the employees. This is especially true in the digital age, where information is key to maintaining a competitive advantage. A recent McKinsey study predicts that by 2028, a large percentage of Chief Analytics Officers will have at least one product they are responsible for, which is primarily based on data and analytics. This takes the use of analytics beyond Davenport's competing on analytics and instead makes it not only competing on analytics but also operating on analytics. In other words, analytics is actually the product or service and is no longer the antecedent of the product or service. This is a huge shift. The author can remember when he first started and companies had to be shown a case for analytics to be able to use it. Today, the firm's product and/or service doesn't function or even exist altogether without it. If this is where we are now heading from a commercial perspective, then everyone in the firm should have some responsibility for producing insights which the firm would need to operate at several levels. In fact, everyone responsible for insights will get firms to this forecasted state of existence faster and enable them to be more competitive.

This author has actually taken it one step further by applying an extended set of operations management techniques, in addition to those associated with supply chain management, to his own data management and analytics projects. One example is the principles of Six Sigma, which will be discussed in detail later in this chapter.

Queuing theory, which is the study of lines and queues, can be applied to Agile projects related to data management software and databases and can deal with backlogs, which are essentially queues. If the data and analytics team is to produce consistent analytics in a timely manner, one may wish to run the creation of insights as a decision engine and use Agile project management to develop and maintain this engine. If this is the case and Agile is being used, then there will be a backlog of items that will form a queue of sorts, with insights ready to be produced. This backlog will include many insights which have been requested by senior management. It's important to manage these queues so that the delays to receiving the insights are not too long, as this would upset the senior executives. In essence, these senior executives are in a queue, waiting for their insights. There could be more senior executives in front of the line, or there could be more urgent requests. Moreover, these executives should have some assurance that their requests are not disappearing into a black hole, which queuing theory can help to mitigate by first understanding the characteristics of the arrival of the requests for analytics, the queue itself, and then how the requests actually get serviced in the queue. It prevents the likes of one of the author's favorite examples: the CEO of a Fortune 500 company asking about the value of a critical partnership two days before Christmas. This required the team to drop all the work they were doing to solve this problem. Of course, it was a high-level request. However, all of the documentation was there to show how this request was going to delay other requests, which were also a priority. In the end, the executive moved to the top of the queue, but the process showed him that his decisions were impacting other insights needed by managers to achieve their goals and objectives. Queuing theory presented to executives like project management, will hopefully make them more comfortable with the wait for their insights and let them know that their insights queue has the right priority.

Process management can help to both visually map the process of how data flows into the systems and models, and process management metrics can measure the ability of data to flow into the process and identify bottlenecks in the data flows. All new data managers should develop process maps, and if they are deploying Lean should add value stream mapping to their data flows. After these maps are developed the next step is to choose a flow metric. The flow metric should always be related to what you are trying to push through the process. In this case, it would be quality data used for insights. Once the process is mapped and the flow metric identified, then process metrics can be used to measure and optimize the process. Remember, it was W. Edwards Deming and others who often stated: "You cannot manage what you don't measure". Once the measurements are in place, improvements can be made. In addition, monitoring can be instituted to ensure drift is prevented and the data quality improvements that were put in place remain in effect.

The author had an incident when he worked for an insurance company that was undergoing a digital transformation. The insurance company CEO had very strong data management practices and even hired an SVP of data management, aka, Chief Data Officer. The organization also had a Chief Analytics Officer, so everything was fairly solid in terms of the data. The problem arose when the company was undergoing a digital transformation. As such, not one person ever looked at the data from the websites because the output was never measured, nor did it need to be measured. By measurement, we mean actually measuring what the website was intended to fulfill (i.e., originating new customers). Of course, web analytics employees had been analyzing things like web clicks and click-through rates using tools like Google Analytics. However, now the ask was something different. Now, we, as an organization, were for the first time evaluating the efficacy of our digital transformation activities through the lens of the website data.

One prime example of where the analytics were needed stemmed from the fact that over $1 billion was spent on phone calls from customers asking about their insurance services. The question management was asking was why this firm couldn't be like Amazon and rarely take phone calls from customers, instead having them rely on the website (including unmanned chats) for their information. This is, no doubt, a legitimate question. Interestingly, this phenomenon is not just isolated to one firm in the sector but is instead endemic across the entire healthcare insurance business. For example, the Blue-Cross Blue Shield health plans are non-profit regional players with an average of a quarter of a million members. Their annual spending on customer service related to inbound calls is well north of $200,000. Most companies would love to discourage people with questions from calling in so they cannot incur the cost. However, the healthcare system in the US is so complicated that the average participant often ends up encouraging members or customers to phone in. By doing this, they can avoid even lower NPS scores than before. Healthcare insurance companies are pretty low on the industry NPS score continuum. Their rank is consistently lower and closely matched to telecommunications firms or cable companies, which, by the way, are all in the 75th percentile. Most insurance plans with commercial plan members at 50+ employees have self-insured plans. In these self-insured plans, the items covered are the same as in a fully insured traditional plan that we are all familiar with (or at least we think we are familiar with). However, in the fully insured plan, the commercial customer (employer) is responsible for paying the claim out of pocket. The only insurance portion could be stop-loss insurance protecting the firm from total claim costs reaching a certain amount. Overall, this is a more affordable and tax-advantaged way for an employer to offer employees healthcare benefits. Given this arrangement, the employer is in the position to approve or deny any claims by employees. Now, they do this through vehicles like drug formularies and explanations of benefits. However, the employee is typically fooled into thinking that it's the payor who is actually making the call on coverage for the employee. This is where the hit is taken on net promoter score, given that the industry is rated at or near the same ranking as telecoms and cable companies, and we all know why these two industries are ranked low, just from our own experiences. So basically, you have a process where the employer is responsible for the decisions, but the service provider takes the hit for denied claims, since the customers are not aware of how the process works.

Well, in order to make a change to this process and reduce the amount of capital directed to customer service, some modifications to the website were needed. And then, after the changes were made, they needed to actually measure the call volumes and see if these adjustments were effective. One can see how this measurement activity is substantially different from the web analytics, which was looking at internal website data only. This activity was now under the moniker of digital analytics and required pulling data from multiple systems. Unfortunately, this was the first time the data was ever analyzed in this way. This was amazing, considering the company had spent millions and millions of dollars on data management and data quality. However, even when there is a mature data management platform, when data is used for a new purpose, the quality may be extremely low since it's never been measured before. It was Dr. William Edwards Deming, among others, who was quoted as saying "You can't manage what you don't measure". This is absolutely the case for new data applications in shops that have great data management and quality control. If the data is not measured properly, then there is no telling the quality of the data. This can be tremendously embarrassing to some "seemingly" well-managed data shops. But it is what it is, and the lesson here to remember is that data quality efforts are only for the data being currently used in the known applications.

3.2 LEAN SIX SIGMA AND DMAIC

Now, if one is going to fix the data quality issues, the best advice at this point is to use a Lean Six Sigma quality improvement paradigm. Lean Six Sigma is the perfect process management improvement technique for addressing data quality issues. It is recommended that one uses the DMAIC (Define, Measure, Analyze, Improve, Control) approach to first define what the quality problems are in the Define Stage. Then, during the Measure stage, the actual extent of the data quality issues are measured. At this stage, we would also benchmark the data quality against certain standards. Then, during the Analyze Phase, we would find the root cause of the situations creating the data quality issues. In the author's experience, it was the web designers who were the root cause of the quality issues, creating issues with tag firing every time they updated the websites. When asked to stop doing this, they claimed that updating the websites in a timely manner was their business goal, and they were not about to slow down. When it came to the Improvement Phase, this is where the improvement plans were implemented and tested. And, finally, during the Control Phase, tools like the FMEA (Failure Mode and Effect Analysis) were used to make sure that the quality improvements didn't slip back to the original level of defects and that the data owners knew how to measure and what to do if there was too much drift back to lower quality. Control Charts from the TQM (Total Quality Management) movement are also useful. The Control phase shouldn't necessarily signal the end of the DMAIC cycle. Leaders can actually immediately propose another DMAIC cycle to implement continuous improvement and focus on some of the items on the Pareto chart that were deprioritized in the first round. This represents the Lean Six Sigma DMAIC solution for addressing and managing poor data quality. Now, of course, there are certainly alternative solutions for measuring and improving data quality. However, the author has found that, given the fact that data is required for effective DMAIC improvement initiatives (the more data the better), and since data is the subject of the improvement here, and all of the phases of DMAIC are important to typical data quality projects, Lean Six Sigma DMAIC is often the best choice. Now, one does not have to be a Lean Six Sigma Black Belt or Master Black Belt (these represent the highest levels of training in Lean Six Sigma) to be able to run a Lean Six Sigma DMAIC data quality improvement project. Participants can pick and choose the tools that are useful for their particular problem. The authors recommend consulting the resources available from the American Society for Quality (ASQ). Their website has been provided in the list of references for this chapter.

Hopefully, this chapter has provided a good overview of how the techniques of operations management can be used to improve data management within a firm. This author strongly recommends that data managers adopt a supply chain management approach to their data management. Data is just as important to firms today as physical goods are in the creation of physical products and needs to be treated in a similar way within the enterprise operation.

REFERENCES

American Society for Quality. n.d. Retrieve from: https://asq.org/

Davenport, T., Evgeniou, T., and Redman, T. 2021. Your Data Supply Chains are Probably a Mess. Here's How to Fix Them. *Harvard Business Review*. pp. 20-25.

Fogarty, D. 2014. From "Silent Spring" to "Noisy Summer": Globalization of industry as a mitigant of domestic environmental problems and the attraction of knowledge workers. *Scholars Journal of Economics. Business and Management* 1(6): 235–242. https://europub.co.uk/articles/-A-384320

Florea, C. 2023, September 17. 20 Ways Oldsmobile Revolutionized the Car Industry. *Autoevolution*. https://www.autoevolution.com/news/20-ways-oldsmobile-revolutionized-the-car-industry-221159.html.

Ford Rubber Plantations in Brazil - The Henry Ford. 2015, October 28. *Thehenryford.org*. https://www.thehenryford.org/collections-and-research/digital-resources/popular-topics/brazilian-rubber-plantations/.

Nyla, P. 2022. The development of synthetic rubber and its significance in World War II. *History in the Making* 15, Article 8. https://scholarworks.lib.csusb.edu/history-in-the-making/vol15/iss1/8

4 Data Quality – Measurement and Its Impact[1]

4.1 INTRODUCTION

With the increasing importance placed on data and analytics in this data-driven world, data capability and analytics capability should be viewed in the same way as other resources such as people, facilities, and raw materials. In various organizations, data management has become a critical function to achieve business success. In this chapter, we will discuss the importance of data quality, concepts, and methods to achieve a high degree of confidence in the data, so we can be confident that the data is fit for its intended function.[2] This helps generate accurate reports, perform high-quality analytics, and make appropriate decisions based on their outcomes.

4.2 DATA QUALITY AND ITS IMPLICATION (THROUGH LOSS FUNCTION LENS)

Dr. Genichi Taguchi, the world-renowned quality expert, highlighted the importance of having good quality products to minimize overall loss. Taguchi (1987) described the relationship between poor quality and overall loss by using a quadratic loss function (QLF) approach. The quality loss function describes the loss that a system produces from a characteristic or attribute such as speed, strength, or time. According to the QLF, the loss increases if the characteristic y deviates from the target (m)—there is a loss associated when the quality characteristic moves away from the target. Taguchi's philosophy terms this loss as a loss to society because someone in society has to pay for this loss. Here, society includes customers, organizations, the government, etc. These losses can have adverse effects, resulting in system breakdowns, accidents, unhappy customers, company bankruptcies, etc. Figure 4.1 shows how the loss increases if the characteristic deviates from the target (on either side) by $\Delta 0$, and is given by L(y). When y equals m, the target value, the loss is zero or at a minimum. The equation for the loss function can be represented as follows:

$$L(y) = k(y - m)^2$$

where, k is a factor based on different types of costs including direct costs, indirect costs, warranty costs, reputational costs, loss due to lost customers and costs associated with rework and rejection.

DOI: 10.1201/9781003190325-4

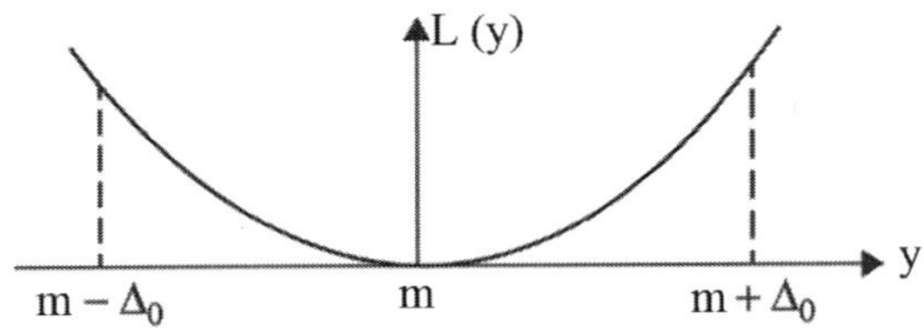

FIGURE 4.1 Quadratic loss function (QLF).

In the data quality (DQ) world, the concept of the loss function can be extended when dealing with the quality of attribute/data elements such as customer social security numbers, customer addresses, and account balances. From the list of data elements, critical data elements (CDEs) are selected based on certain rationalization criteria. The data quality of these CDEs is typically measured in terms of percentages. These percentages are based on dimensional scores related to data accuracy, data completeness, data conformity, and data validity.

If the DQ levels associated with these CDEs are not on or close to target values, there is a high probability of making wrong decisions that could have adverse effects on organizations. Since the DQ is measured in percentage, higher values are desirable. So, in this case, only half of QLF is applicable when measuring loss due to poor data quality. The loss function while addressing of data quality is shown in Figure 4.2. Using this figure, we can see how the overall loss increases if the DQ levels (of a CDE) is away from the target, m. Sometimes in DQ context, the target values are also referred to as thresholds. As we can see in Figure 4.2, the loss will be at minimum when y reaches the target level, m.

Poor data quality may incur losses in several forms. Taguchi (1987) classifies the effect of poor quality into two categories: losses caused by the functional variability of the process and losses caused by harmful side effects. Figure 4.3 shows how all these costs accumulate, and one can imagine how they will add up to cause an overall loss to society.

The importance of ensuring high-quality data was emphasized by famous statisticians long before the data field experienced massive growth. World-renowned statistician, Dr. C.R. Rao (1989) provided a thorough checklist for vetting data. Some of the questions he recommends asking are listed below.

C.R. Rao's Checklist:

- How data is collected and recorded?
- Is the measurement system well defined?
- Is the data free from recording errors?

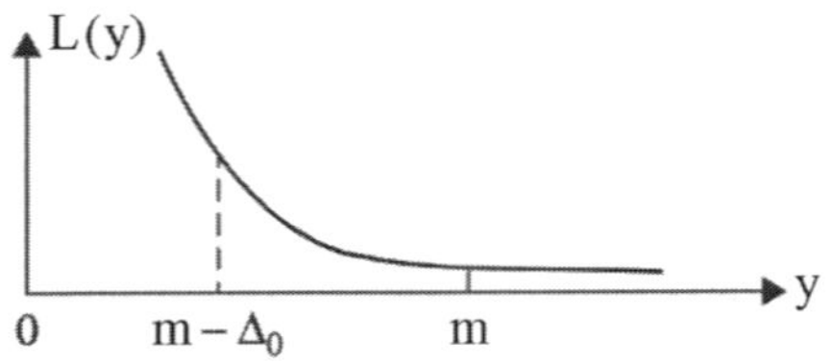

FIGURE 4.2 Loss function in the context of data quality.

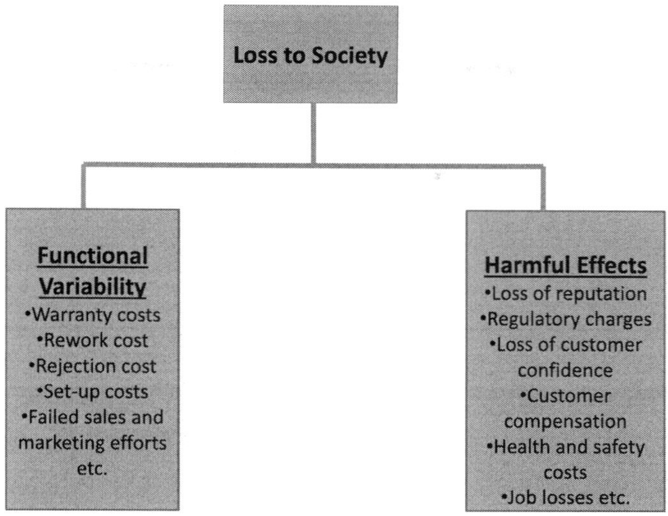

FIGURE 4.3 Loss to society sources.

- Is the data from reliable sources and trustworthy?
- Are there any abnormalities or outliers associated with data?
- Is sample size adequate? Are all factors for data collection considered?
- Do you have the right kind of data that is suitable for the intended purpose?

4.3 THE DATA QUALITY OPERATING MODEL[3]

For the data to be fit-for-purpose, there are two questions that need to be addressed: 1. Do you have the right data? 2. Is the data right? The first question can be addressed by clearly defining the business problem we are trying to solve. A good data quality program should address the second question. This program should cover the entire organization's data. To ensure high-quality data, there are several requirements that must be satisfied. For this, we should have a structured data quality (DQ) program that can be applied across the organization. The overall data quality program should drive business value so that organizations can be competitive based on reliable data.

4.3.1 Data Quality Capabilities

A typical DQ program will focus on the six core capabilities as shown in Figure 4.4

1. **Strategy and Governance:** This includes understanding the current state of data quality for critical data and the gaps compared to the target. The focus will also be on how to improve DQ by reducing the gap so that enterprise strategic goals can be met.
2. **DQ Resources**: DQ resources refer to highly-skilled people who can execute the DQ program by understating the data and associated processes.

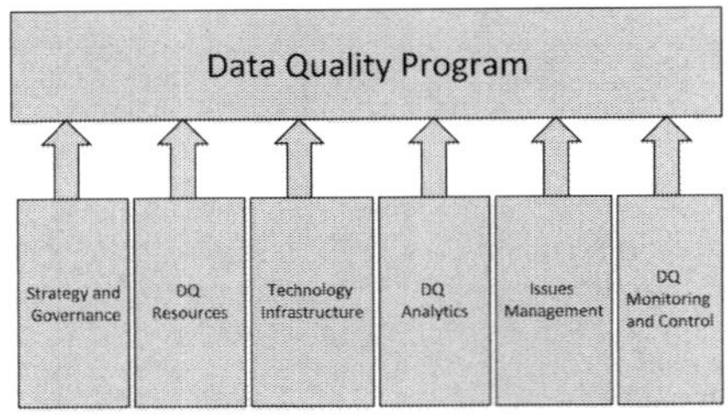

FIGURE 4.4 Data quality capabilities.

3. **Technology Infrastructure:** This focuses tool infrastructure, and platforms required to collect, manage and analyze the data through appropriate concepts/methods. There should also be a capability for producing reports based on information related to critical organization's data.
4. **DQ Analytics:** This capability should focus on the processes, techniques and tools needed to measure DQ levels of critical data. Further, there should also be a framework to conduct root cause analysis for resolving data issues.
5. **Issues Management:** Issues management process should focus on identification, triaging, tracking, and resolving of data quality issues. Root cause analysis and remediation efforts are very important for issue management process.
6. **DQ Monitoring and Control:** DQ monitoring and control are continuous activities to measure and control DQ levels as needed. To do this, there is a need to establish a monitoring and control environment, as well as a change management process. Typically, statistical process control (SPC) charts are used in monitoring and controlling efforts.

For implementing these core data quality capabilities, we should have a structured approach.

4.4 STRUCTURED APPROACH FOR SOLVING DATA QUALITY PROBLEMS (DAIC APPROACH)

In this section, we will describe a structured approach with four phases to solve data quality problems. This is a phased approach that is carried out in four phases – Define, Assess, Improve, and Control. This approach sometimes is referred to as the DAIC approach. This holistic approach is aimed at improving data quality and sustaining the gains made in the Improve phase.

This approach is built similarly to Six Sigma phased approaches to improve process performance levels. The DAIC methodology and its phases are illustrated below with the help of Figure 4.5.

4.4.1 The Define Phase

In the Define phase, we focus on problem definition and establishing the scope, objectives, resources needed, and good project plans. The most important activity in the Define phase is the creation of the project charter, which helps establish the

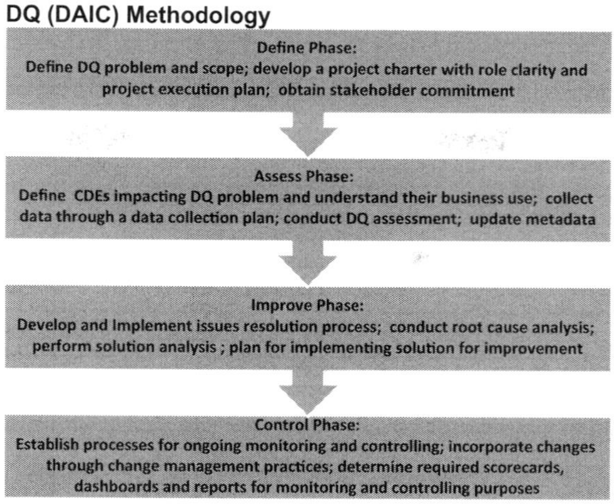

FIGURE 4.5 DAIC based DQ methodology.

scope, objectives, resources, expected business value, and role clarity. The project managers need to prepare a detailed project plan for the four phases of DAIC, with the relevant tasks and key deliverables.

4.4.2 The Assess Phase

The Assess phase focuses on defining the business use for the critical data, establishing a data quality baseline, and assessing the current state. In this phase, it will be important to narrow the focus of assessment and monitoring to critical data. Critical data is required to deliver the key business outcomes and deliverables. Since data quality is measured based on data elements, it is necessary to identify critical data elements (CDEs). Now, let us formally define data elements and critical data elements.

Data Element: Data elements can be defined as data characteristics that are required in business activities. The data elements can take varying values. Examples of data elements are social security numbers, customer IDs, date of birth, account balance, type of facility, commercial bank branch, etc.

Critical Data Element (CDE): A critical data element can be defined as a data characteristic that is "critical" for the organization to conduct business through analytics or reporting activities. Typically, CDEs are identified for different domains of the business. Examples of domains include risk, compliance, claims, customer service, etc.

Jugulum (2014) provides an approach for identifying critical data elements through rationalization analysis. To identify CDEs, a sampling-based statistical analysis called the funnel methodology is recommended. By applying the funnel methodology, we can reduce the number of data elements by using a rationalization matrix, correlation and regression analyses if the data is of a continuous type, and association analysis if the data is of a discrete type. This helps us identify CDEs that have close relationships. After this, a signal-to-noise ratios (S/N ratios) analysis is conducted for each pair of highly correlated CDEs. The CDEs with lower S/N ratios are chosen for future assessment, as a lower S/N ratio indicates higher variation. Higher variation indicates that the process generating the data is not stable and might impact the smooth functioning of the business. Figure 4.6 illustrates the CDE identification approach.

In Figure 4.6, the approach starts with a set of data elements that were identified based on inputs from subject matter experts. An example of funneling data elements to determine CDEs is shown on the right side of this figure. Let us suppose that we have 100 data elements before applying the funnel methodology. After performing rationalization analysis and a prioritization method like Pareto analysis or any other method, we could reduce the number of CDEs to 60. After applying statistical tools such as correlation and association analysis, this number can be reduced to 30. Then, through S/N ratio analysis, this can further be reduced to a manageable number of 15. These 15 data elements are CDEs that are important, and they are qualified for DQ assessment. DQ assessment is generally performed by using data profiling techniques that include basic analysis of missing values, validity against known reference sources, accuracy and data formatting checks, and descriptive statistics to understand distributional aspects or patterns associated with CDEs. The assessment also includes measuring the gap between current DQ levels and thresholds.

Data Quality Dimensions: Data quality dimensions are associated with CDEs or data elements, and DQ scores are computed based on the DQ dimensions. Wang and Strong (1996) define data quality dimensions as a set of attributes that represent a single aspect of data quality. The following four data quality dimensions are commonly used and evaluated by various rules against data sets to compute DQ.

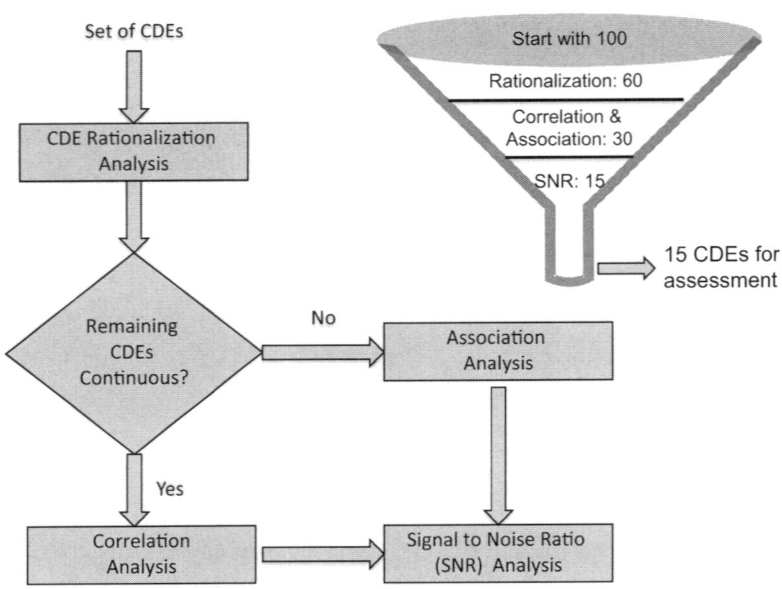

FIGURE 4.6 CDE identification through the funnel approach (with illustrative example).

- Completeness
- Conformity
- Validity
- Accuracy

Typically, we calculate DQ scores at these dimensional levels against business rules. Then dimensional level scores are used to calculate DQ scores for CDEs.

Table 4.1 shows descriptions of these four data quality dimensions.

4.4.2.1 Measuring Data Quality

To measure the data quality (DQ) levels for CDEs, we need to select the data quality dimensions relevant to the specific business process. Typically, the four DQ dimensions described in Table 4.1 are used. After selecting DQ dimensions, business rules are applied to obtain dimensional level scores. By applying the data quality rules to critical data elements (CDEs), we classify them as acceptable or unacceptable in the context of a chosen dimension.

4.4.2.2 Measurement of Data Quality Scores

After developing rules corresponding to the data quality dimensions and validating them against these rules, data quality scores (DQ scores) are calculated. DQ scores measure data performance and indicate if the data is fit-for-purpose. A DQ score can be a measure for a given data quality dimension, an aggregated score of multiple data quality dimensions of a CDE, or even an aggregated score of multiple CDEs at either the function/business unit level or at the enterprise level. DQ scores are usually measured in percentages.

Data quality scores at multiple levels are usually calculated in a sequential manner. They are measured at the dimensional level first, then rolled up at the CDE level, followed by the taxonomy level, and then at the function/business unit level. After this step, enterprise-level DQ scores are calculated. Figure 4.7 shows this sequence of calculations to be performed.

TABLE 4.1
Commonly Used DQ Dimensions

Dimension	Definition
Completeness	Completeness is a measure of the presence of core data elements that is required in order to complete a given business process.
Conformity	Conformity is defined as a measure of a data element's adherence to required formats (data types, field lengths etc.)
Validity	Validity is defined as the extent to which data corresponds to valid values as defined in authoritative sources.
Accuracy	Accuracy is defined as a measure of correctness of the data element as viewed by a valid real world source.

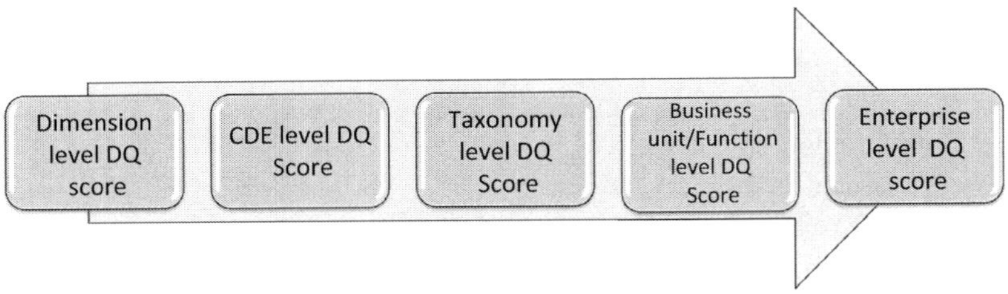

FIGURE 4.7 Measuring enterprise level DQ score.

4.4.2.3 Measuring Enterprise Level DQ Score

Figure 4.8 describes the process of obtaining DQ scores at various levels. The approach in Figure 4.8 is based on giving equal weights to all dimensions, CDEs, taxonomies, and functions. If weights are to be used, then weighted averages can be applied.

4.4.3 The Improve Phase

After assessing DQ in the Assess phase, the DAIC approach focuses on DQ improvement activities. The issues causing lower DQ scores need to be identified, and root cause analysis must be performed. The issue management system should capture and prioritize these issues. The data, technology, and business teams should work

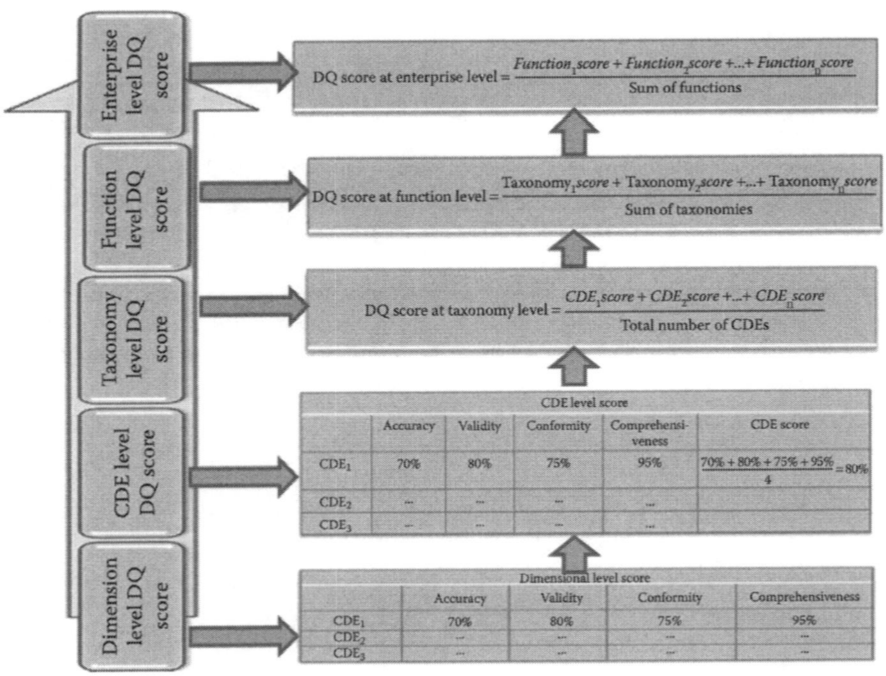

FIGURE 4.8 Data quality scores at various levels.

together to conduct root cause analysis on the issues and resolve them. Remediation and improvement efforts must be deployed and constantly monitored to ensure visibility of DQ at different levels of leadership.

4.4.4 The Control Phase

The final phase of the DAIC approach is the Control phase, which aims at monitoring and controlling the improvement activities with scorecards, dashboards, etc. Usually, ongoing monitoring and improvement activities would include statistical process control aspects for monitoring DQ performance.

In summary, the DAIC approach will help ensure data is fit for intended purpose by gaining control on the key data used in business processes so that effective decisions are made.

4.5 DATA QUALITY IMPROVEMENT PRACTICE USING CDE FUNNEL APPROACH[4]

A large bank established an enterprise-level data management function with the objective of continuously improving data quality and consistency across the bank's various functions and business units. This function drives the data governance discipline and culture so that high-quality data and information are ensured to successfully support the bank's operations. This function also incorporated DQ measurement in its continuous efforts to identify, assess, and manage all types of risk – regulatory, liquidity, credit, market, operational, etc. Stronger DQ capabilities are necessary, as they allow organizations to increase customer and regulator confidence in the data that supports products, services, processes, and reporting. In addition, these capabilities enable organizations to increase the effectiveness and efficiency of all operations.

As we know, the data quality assessment and improvement begin with identifying which critical data elements (CDEs) are to be monitored, assessed, and improved. To identify CDEs, a funnel methodology was developed so that the right CDEs are selected from a large pool of data elements. The funnel approach is described in Chapter 2. This particular study is related to the Basel II accords/standards. Financial companies have started adopting the regulatory measures outlined in the Basel II Accords. These accords are designed to ensure that banks follow a disciplined process to set aside sufficient capital to offset the risks created through their operational practices.

4.5.1 CDE Rationalization Matrix

At the beginning of this effort, the risk function of this bank started with a list of 35 data elements, from which CDEs would be selected. A CDE rationalization analysis was used for prioritizing data elements and selecting CDEs. Prioritization was done by scoring the CDE dependencies against weighted ranking criteria and deriving a total, which is a sum product of all criteria weights and their corresponding scores. This tool helped in conducting a high-level potential-CDE filtering based on science and experience. The range of total scores (223–595) provided a good means

Weightages / Criteria / CDEs	10	7	7	7	10	10	10	7	7	4		
	Ease of access	% of reports	% of customers	Total incidences	Business support	Technolgy support	Regulatory risk	Financial risk	Reputation risk	Operational risk	Total Score	
CDE1	7	10	7	10	1	7	4	7	4	7	484	
CDE2	1	10	7	7	1	4	7	10	4	10	436	
CDE3	4	7	1	10	10	4	4	1	1	7	388	
CDE4	10	7	1	1	1	4	7	7	10	4	418	
.....	...	...	...	...	...	...	...	...	...	...	...	
CDE22	4	4	4	10	10	10	7	10	7	10	595	Highest
CDE23	4	4	1	4	1	10	4	4	7	1	292	
....	...	...	...	...	...	...	...	...	...	...	...	
CDE35	4	10	7	7	1	7	1	4	1	7	361	

FIGURE 4.9 CDE rationalization analysis matrix.

to decide on the criticality of data elements, besides proving the value of the tool. Through this approach, the number of CDE candidates under consideration was reduced from 35 to 24. Figure 4.9 shows a sample CDE rationalization and prioritization analysis matrix.

4.5.2 Correlation and Association Analysis

In the next step of prioritization, statistical correlation and association analysis techniques were carried out on the information related to the 21 remaining candidate CDEs with the intent of capturing the relationships and dependencies between CDE candidates. In cases where strong correlations exist between two candidate CDEs, we can keep one of the two CDE candidates. Similarly, in the case of attribute-type data element variables, dependencies can be established through the measure of association. In this case, CDE 5 and CDE 8 were of attribute type, and an association analysis was performed to determine the relationship between these two potential CDEs. It was found that there was no relationship between CDE 5 and CDE 8, prompting the risk team to keep both of them for data quality assessment.

For the remaining 19 CDEs, a correlation analysis was performed. In the correlation analysis Table 4.2 (which is shown partly), there were eight instances (pairs) of strong correlations with correlation coefficients greater than 0.85. These pairs were: CDE4-CDE7, CDE4-CDE10, CDE4-CDE12, CDE10-CDE7, CDE10-CDE12, CDE15-CDE16, CDE13-CDE33, and CDE28-CDE29. In the final step of the prioritization process, it was required to determine which of the candidate CDEs in a correlated pair should be kept for DQ assessment. This determination was made through the calculation of the signal-to-noise (S/N) ratios.

4.5.3 Signal-to-Noise Ratios (S/N Ratios)

Through S/N ratio analysis, one can choose which CDE in a related pair needs to be kept for data quality assessment. A higher S/N ratio means lower variability. So, CDEs with low S/N ratios require DQ assessment in the form of monitoring and control. Table 4.3 shows the S/N ratio analysis for highly correlated CDEs.

TABLE 4.2
Correlation Analysis

	CDE4	CDE6	CDE7	CDE9	CDE10	CDE12	CDE13	CDE15	…	CDE24	CDE27	CDE28	CDE29	CDE30	CDE33
CDE4	1	−0.158	0.941	−0.070	1.000	0.867	−0.002	−0.048	…	−0.027	0.121	0.057	0.331	0.105	−0.002
CDE6	−0.158	1	−0.136	0.094	−0.158	−0.157	−0.171	0.034	…	−0.012	0.277	0.050	−0.041	0.276	−0.171
CDE7	0.941	−0.136	1	−0.070	0.941	0.822	−0.025	−0.044	…	−0.025	0.128	0.044	0.297	0.187	−0.025
CDE9	−0.070	0.094	−0.070	1	−0.070	−0.066	−0.039	0.015	…	−0.012	0.077	−0.016	−0.050	−0.034	−0.039
CDE10	1.000	−0.158	0.941	−0.070	1	0.867	−0.002	−0.048	…	−0.027	0.121	0.057	0.331	0.105	−0.002
CDE12	0.867	−0.157	0.822	−0.066	0.867	1	−0.002	−0.048	…	−0.018	0.122	0.046	0.303	0.087	0.039
CDE13	−0.002	−0.171	−0.025	−0.039	−0.002	0.039	1	0.02	…	0.03	−0.24	−0.22	−0.17	0.04	1.00
CDE15	−0.048	0.034	−0.044	0.015	−0.048	−0.048	0.02	1	…	0.000	−0.013	−0.031	−0.028	0.020	0.020
…	…	…	…	…	…	…	…	…	…	…	…	…	…	…	…
CDE24	−0.027	−0.012	−0.025	−0.012	−0.027	−0.018	0.03	0.000	…	1	−0.038	−0.033	−0.035	−0.043	0.033
CDE27	0.121	0.277	0.128	0.077	0.121	0.122	−0.24	−0.013	…	−0.03828	1	−0.03218	0.1152	0.1743	−0.23989
CDE28	0.057	0.050	0.044	−0.016	0.057	0.046	−0.22	−0.031	…	−0.03277	−0.032	1	0.867	0.132	−0.218
CDE29	0.331	−0.041	0.297	−0.050	0.331	0.303	−0.17	−0.028	…	−0.03514	0.115	0.867	1	0.106	−0.170
CDE30	0.105	0.276	0.187	−0.034	0.105	0.087	0.04	0.020	…	−0.04288	0.174	0.132	0.106	1	0.044
CDE33	−0.002	−0.171	−0.025	−0.039	−0.002	0.039	1.00	0.020	…	0.03347	−0.240	−0.218	−0.170	0.044	1

Through the correlation matrix (Table 4.2), we can see that CDE 4, CDE 7, CDE 10, and CDE 12 have mutually high correlations. CDE 7, with an S/N ratio of 6.27 decibels, has the lowest S/N ratio compared to the others. For this reason, CDE 7 was selected for data quality assessment. Similarly, CDE 15 was chosen from the pair CDE 15 and CDE 16. For the pair CDE 28 and CDE 29, CDE 29 was selected. Since CDE 13 and CDE 33 had the same S/N ratio, business reasons were applied for the selection, and accordingly, CDE 13 was selected for DQ assessment.

So, after the final steps, 15 of the original 35 data elements were selected as the Critical Data Elements, and DQ assessment was conducted on them. Figure 4.10 shows the Funnel Approach of reducing the CDEs from 35 to 15.

Table 4.4 shows a summary of the DQ assessment corresponding to the CDEs.

TABLE 4.3
S/N Ratio Analysis for Highly Correlated Variables

CDE	S/N Ratio
CDE4	6.42
CDE7	6.27
CDE10	6.47
CDE12	6.54
CDE13	11.96
CDE15	−18.66
CDE16	−18.57
CDE28	−4.65
CDE29	−6.03
CDE33	11.96

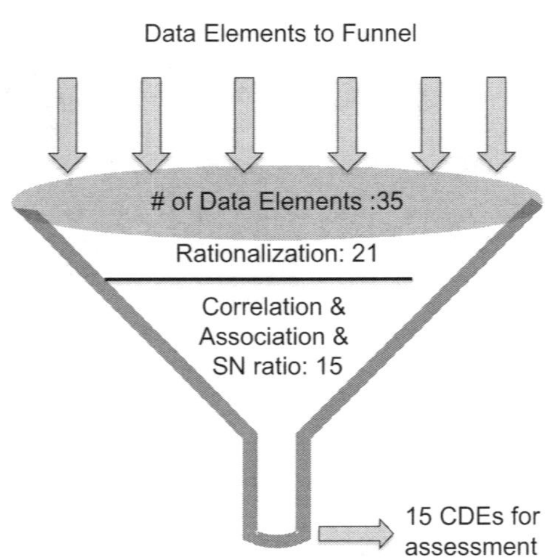

FIGURE 4.10 Funnel approach of reducing the CDEs from 35 to 15.

TABLE 4.4
DQ Assessment Summary for Final 15 CDEs

Dimension	Completeness	Conformity	Validity
CDE			
CDE5	100.00%	100.00%	100.00%
CDE6	100.00%	100.00%	100.00%
CDE7	100.00%	100.00%	100.00%
CDE8	100.00%	100.00%	100.00%
CDE9	99.80%	99.20%	99.28%
CDE13	0.20%	0.00%	0.00%
CDE15	84.40%	86.60%	86.87%
CDE18	100.00%	100.00%	100.00%
CDE19	100.00%	100.00%	100.00%
CDE20	98.40%	98.00%	98.20%
CDE23	0.60%	1.40%	0.36%
CDE24	100.00%	100.00%	100.00%
CDE27	39.00%	38.00%	32.01%
CDE29	100.00%	100.00%	100.00%
CDE30	100.00%	100.00%	100.00%

4.5.4 Impact on the Analytics Quality

The methodology helped in reducing the number of data elements by selecting the vital few. DQ monitoring and controlling were carried out on these 15 CDEs. These 15 critical data elements are typically used in different risk calculators/models to make important decisions for the clients or customers.

4.6 DQ MONITORING AND CONTROL THROUGH SPC TO ENSURE PROCESS QUALITY[5]

This case study focuses on applying SPC in monitoring DQ levels of eligibility data related to processing claims in a health care organization. Eligibility data is composed of demographic information for each customer who is eligible to receive medical, pharmacy, dental, or other insurance coverage and shows if a person can receive benefits with a defined start and end date at a point in time. In the U.S. health care system, employers (clients) purchase most benefits for their employees (customers). Since most clients have a relatively stable number of employees, the number of customers eligible to receive the benefits (called eligibility counts) is also expected to be stable from month to month, with the exception of the enrollment and renewal periods, when employees have the option to change their benefits. The purpose of using SPC on eligibility counts is to proactively determine abnormalities in eligibility data and identify the associated data issues. Developing DQ controls directly into applications helps in the quick identification and resolution of data issues.

4.6.1 Study Details

The counts of eligible people can be gathered at different levels (segments, accounts, and products) because they are expected to be consistent at each of these levels. Here, a segment represents the highest level and corresponds to a grouping of clients in different locations. Accounts represent different clients, and products correspond to coverage and programs for sets of customers within a client organization. An example of a product could be the "healthy life program," where health coaches from an insurance company can advise customers on food habits, exercise benefits, etc. Measurements at the product level within a client are very granular; measuring them produces large volumes of measurement data. So, this is a nested type of design, as shown in Figure 7.9. Since eligibility data flows are complex, SPC-based controls were applied to segments, accounts within segments, and products within accounts. Measurements at each level help identify problems at that level. This set of measurements also helps us understand the impact of downstream problems at higher levels (such as client or segment), making it easier to isolate and resolve issues. Figure 4.11 provides a generalized, nested view with n segments, k accounts in segments, and m products in accounts. In our analysis, we had only three segments and a specific number of accounts and products for each segment.

Here, we present the SPC analysis discussion for one sample account and one sample product within that account. Figures 4.12 and 4.13 show examples of I-charts (individual charts) for this account and product. Since the metric here is eligibility counts, an I-chart is useful, and counts should be between the lower and upper control limits of this chart. The upper control limit is three standard deviations from the mean on the positive side, and the lower control limit is three standard deviations from the mean on the negative side. These limits serve as thresholds or specifications, and for an I-chart, they are calculated from moving ranges. Here, a moving range is the absolute difference between two successive observations. The equations to calculate the upper and lower control limits are as follows:

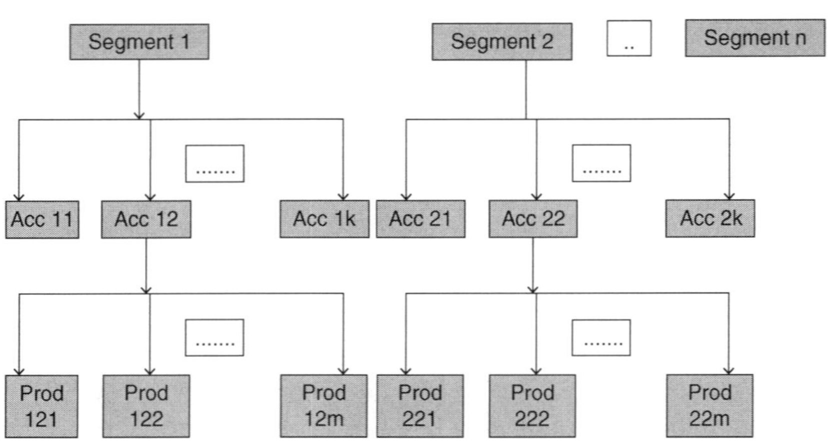

FIGURE 4.11 Nested design for segments, accounts and products.

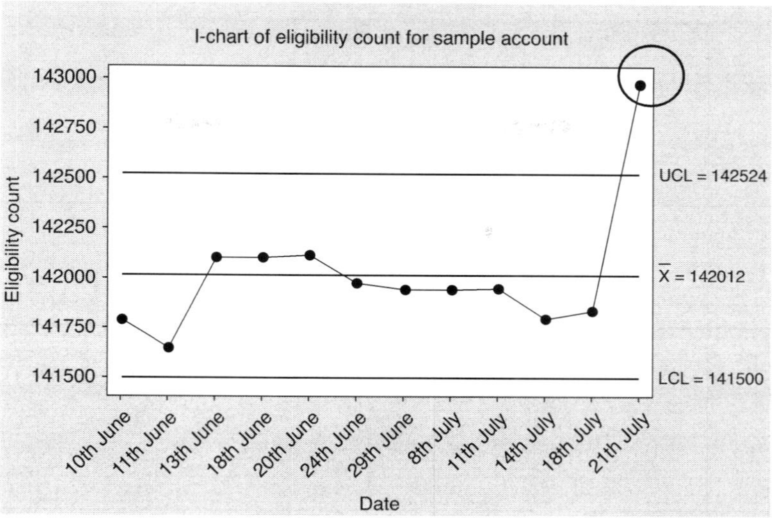

FIGURE 4.12 Thresholds and SPC analysis for eligibility count per account.

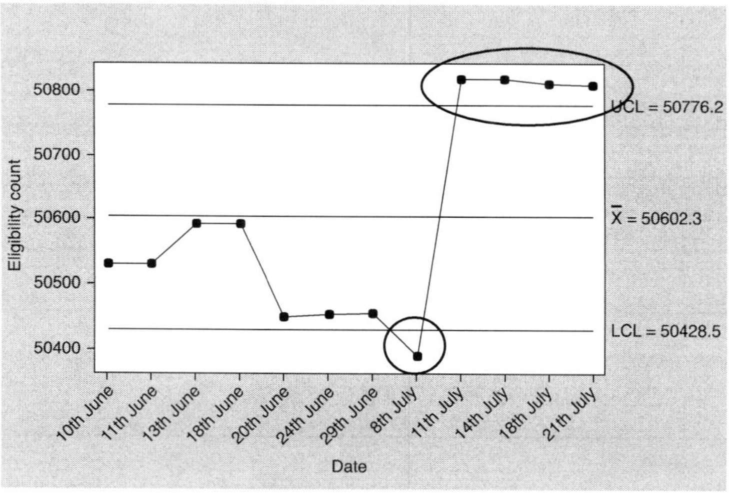

FIGURE 4.13 Thresholds and SPC analysis for counts per day per product group.

$$\text{Upper Control Limit (UCL)} = \bar{X} + 3 \times \frac{\overline{MR}}{4.128}$$

$$\text{Lower Control Limit (LCL)} = \bar{X} - 3 \times \frac{\overline{MR}}{4.128}$$

where:

$\bar{X}$ is the mean of daily eligibility counts
$\overline{MR}$ is the mean of all moving ranges

4.6.2 Out of Control Situations

Figure 4.12 shows an overall increase in the count at the account level, and Figure 4.13 shows an increase at the particular product level. The eligibility counts are monitored against control limits or thresholds. As mentioned earlier, discrepancies in the counts are expected during the periods of enrollment and renewal, which typically occur at the beginning of the calendar year, and changes in other periods are not acceptable. Since discrepancies occurred in July, it was necessary to investigate the issue by identifying the root cause.

Root Cause Identification and Remediation: These out-of-control situations were discussed with all relevant individuals, and the reason for the out-of-control situations was identified: a technical problem in the logic of the program at both the account and product levels. This problem was causing incorrect effective and termination dates. The logic of the program was changed to ensure that the dates are accurate and the counts are correct. This change brought the counts back into statistical control.

Control charts help detect out-of-control situations, as explained above. They also help in cases where downstream out-of-control situations impact upstream levels (meaning that product-level out-of-control situations can impact account-level, and account-level out-of-control situations can impact segment-level). The use of SPC-based control mechanisms can identify the root causes by drilling down to the product level. This is illustrated in Figure 4.14.

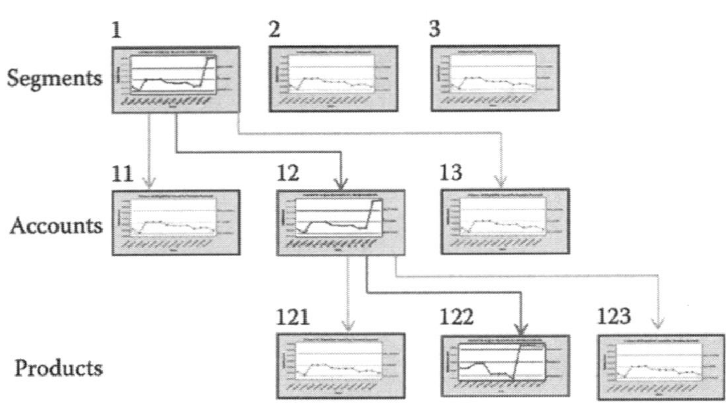

FIGURE 4.14 Tracing root causes through drill-down of SPC charts to product level.

4.6.3 Impact on Process Quality

If the eligibility counts change significantly at other times, it is likely that something is wrong with the business processes. Incorrect counts have an adverse impact on interactions between the insurance company and its customers and clients. Information related to claims payments depends on having complete and reliable eligibility data. Therefore, incorrect or incomplete eligibility data can adversely affect the corresponding processes and result in customer dissatisfaction, potentially causing the company to lose credibility.

NOTES

1 Thanks are due to CRC Press for allowing us to use materials from prior publications of one of the authors.
2 Thanks are due to Wiley for allowing us to use materials from prior publications of one of the authors.
3 The author sincerely thank Brian Bramson for his involvement during the development of DAIC effort.
4 The author worked with C. Shi, H. I. Joyce, J. Singh, B. Granese, R. Ramachandran, D. Gray, C. H. Heien and J. R. Talburt on effort and sincerely thank them. Thanks are due to Wiley for granting permission.
5 Two authors worked with L. Sebastian-Coleman, C. Heien, R. Vadlamudi and D. Gray on effort and sincerely thank them. Special thanks to Laura for providing materials from her publication.

REFERENCES

C. Shi, R. Jugulum, H. I. Joyce, J. Singh, B. Granese, R. Ramachandran, D. Gray, C. H. Heien, J. R. Talburt. 2015. Improving Financial Services Data Quality - A Financial Company Practice. *International Journal of Lean Six Sigma*, Vol 6(2).

Jugulum, R. 2014. *Competing with High Quality Data: Concepts, Tools and Techniques for Building a Successful Approach to Data Quality*. Wiley publication.

L. Sebastian-Coleman, R. Jugulum, C. Heien, R. Vadlamudi and D. Gray. 2016. Statistical Process Control and its Relevance in Data Quality Monitoring and Reporting: IQ *International Journal*, Vol.11 (1).

Rao, C. R. 1997. *Statistics And Truth: Putting Chance To Work*, Wspc publications

Redman, T. C. 1996. *Data quality for the Information Age*. Artech House.

Taguchi, G. 1987. *System of Experimental Design*, Vols. 1 and 2, ASI & Quality Resources, White Plains, New York.

Wang, R. Y. and Strong, D. M. 1996. Beyond accuracy: What data quality means to data consumers. *Journal of Management Information Systems*. 12(4): 5–33.

5 Analytics Landscape, Execution, and Evaluation[1]

5.1 INTRODUCTION

In this data-driven world, we know how important data capability and analytics capability are to make accurate decisions by way of diagnostics or predictions. The quality of analytics is as important as the quality of data. There should be a structured approach/process to perform and govern analytics to ensure the quality of analytics. Such analytics should also be scaled to get repeatable and consistent results. In this chapter, we will discuss the importance of having such a process to ensure both data quality and analytics quality levels are met to enable one to perform high-quality analytics and make appropriate decisions.

5.2 ANALYTICS REQUIREMENTS

For the successful execution of analytics, across any organization, a clear "analytics vision" that includes what will be accomplished, is necessary. This vision should be supported by an "analytics strategy" as shown in Figure 5.1. A good strategy outlines various aspects of how to make analytics work successfully. After defining the vision and outlining a suitable strategy, it is necessary to identify sound analytical talent and resources for the successful deployment of analytics strategy.

After fulfilling the talent and resource requirements, the next step is to make sure that the right kind of data capabilities and analytics platform exist to support various types of data on a large scale. A suitable analytics platform with relevant features such as data ingestion, storage, analytics (including AI and ML), reporting,

FIGURE 5.1 Successful analytics execution.

DOI: 10.1201/9781003190325-5

and monitoring is essential for processing large amounts of data. To systematically expand the use of analytics, it is important to have a standardized approach with an appropriate operating model that will serve as a reference guide for executing analytics, including descriptions of different types of analytics, how to deploy them, how to interpret the results, and associated methodologies. Standardization through operating model will enable analytics adoption throughout the organization. After satisfying analytics requirements, the next step is to define a process for execution.

5.3 PROCESS OF EXECUTING ANALYTICS

Figure 5.2 outlines a seven-step process for analytics execution for any given problem. The first step in the execution process is to define the real-life problem by understanding its purpose. The second step in the process is to formulate the analytical problem by taking real-life considerations into account. The third and fourth steps focus on collecting relevant data and ensuring its quality. After ensuring high data quality in the fourth step, the emphasis will be on deciding the type of analytics we need to use (Table 5.1). The fifth step will focus on inferential analysis by choosing an analytical framework or model. The sixth step involves comparing model outputs with actual results. For a good framework/model, the gap between these two should be small. If the model provides the expected results (that is, when the gap is small), then in the next step, we should focus on the deployment of model outputs

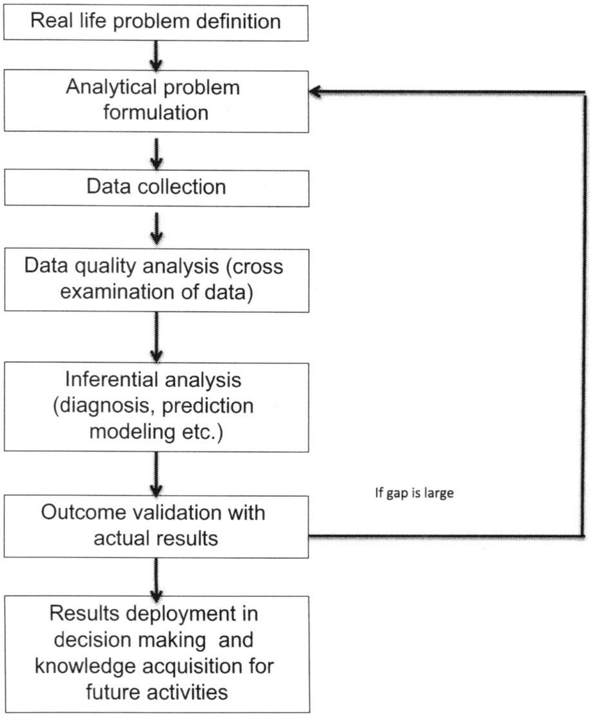

FIGURE 5.2 Analytics execution.

TABLE 5.1
Tools and Techniques for Analytics

Type of Analytics	Tools/Techniques
Preparatory analytics	Simple profiling with DQ rules, statistical process control (SPC) charts and DQ funnel approach etc..
Descriptive analytics	Data profiling ,data mining, descriptive statistics, capability analysis etc..
Diagnostic analytics	SPC charts, analysis of variance, correlation and association analysis, hypothesis tests etc..
Cause related analytics	Cause and effect analysis, failure mode effect analysis, parameter diagram (p-diagram) etc.,
Predictive analytics	Generalized linear models, regression analysis etc.,
Prescriptive analytics	Experimental design, simulation based scenario planning etc..
Reliability based analytics	Signal to noise ratio analysis, failure analysis, confidence intervals, etc.

in decision-making and future knowledge acquisition activities. If the gap between model outputs and actual results is large, then we need to go back to the second step to refine the model and follow the subsequent steps.

Note that in this approach, step four, "data quality analysis", is quite important because high quality data is critical for the successful execution of analytics to achieve meaningful business outcomes. If the quality of data is not satisfactory, the analytics results will not be valid, which can result in a huge loss to society.

The fifth step of the analytics execution process highlights the importance of choosing the right type of analytics for inferential analysis. There are different types of analytics, and they can be selected depending on the problem that one is dealing with. Figure 5.3 shows analytics taxonomy along with their intended purpose. It includes the types proposed by Gartner and those added by Jugulum (2014). Table 5.1 shows the tools and techniques required for these analytics.

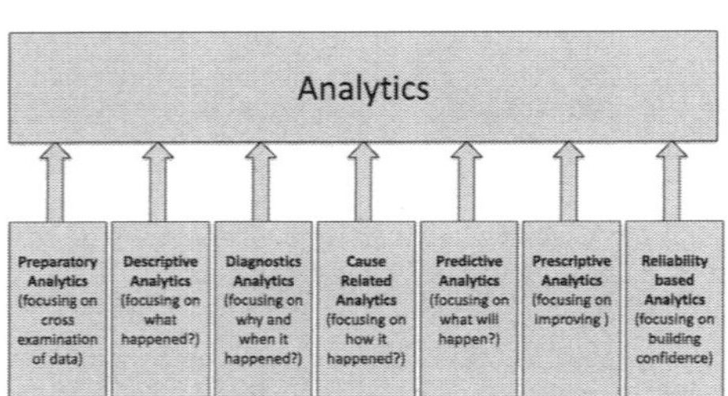

FIGURE 5.3 Taxonomy of analytics.

Brief descriptions of the types of analytics shown in Table 5.1 are provided below:

Preparatory Analytics: The purpose of preparatory analytics is to conduct cross-examination of data to ensure data quality. Typically, simple data profiling techniques, DQ funnel approach, and statistical process control methods are used in this type of analytics.

Descriptive Analytics: Descriptive analytics deals with answering questions like what happened to the performance of a particular CDE/factor, system, process, or product. Data profiling and data mining tools, as well as stability and capability analysis using descriptive statistics, are quite useful in this type of analytics.

Diagnostic Analytics: This type of analytics is useful for understanding when, where, why, and how a particular problem has occurred. Typical tools and techniques used in this type of analytics include: correlation analysis, association analysis, hypothesis tests, analysis of variance (ANOVA), and control charts.

Cause-related Analytics: Cause-related analytics are intended to conduct root cause analysis for understanding the causes of problems or failures. Ishikawa (cause-and-effect) diagrams, cause-and-effect matrices, parameter diagrams (P-diagrams) and failure mode and effects analysis (FMEA) are quite useful for this type of analytics.

Predictive Analytics: If the goal is to predict the behavioral patterns associated with a process, product, system, or CDE, predictive analytics are used. Tools and techniques like general linear models and regression analysis are useful for performing predictive analytics.

Prescriptive Analytics: Prescriptive analytics is useful for answering questions like how we can improve performance. Tools like experimental designs are very useful here. Simulation based scenario analysis is also used in prescriptive analytics.

Reliability Based Analytics: To estimate the reliability of a product, process or system, or set of models with a required degree of confidence, we use this type of analytics. Failure analysis, confidence intervals, and signal-to-noise ratios etc. are quite useful in this type of analytics.

5.4 OVERALL ANALYTICS EXECUTION PROCESS IN DMAIC PHASES

The goal of the data analytics process should be to aid organizations in handling large amounts of data to make better business decisions by obtaining better outcomes from analytics.

Processing data and performing efficient analytics require high-end technologies/platforms and adequate resources. Along with these, there should be a structured approach to performing analytics. The six Sigma-based DMAIC (Define, Measure, Analyze, Improve, and Control) framework is also well suited for executing the overall analytics process. Figure 5.4 explains DMAIC based analytics execution.

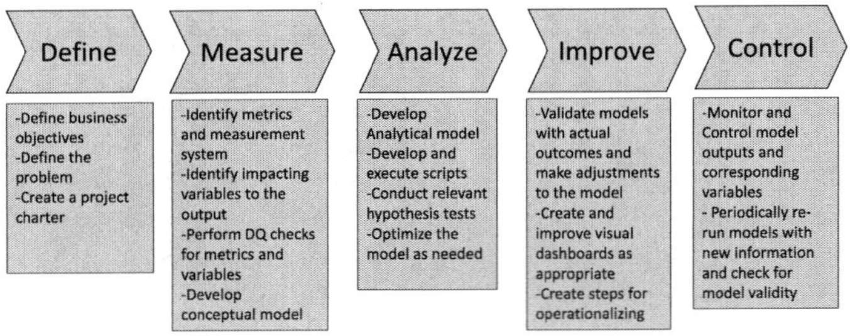

FIGURE 5.4 Structured DMAIC approach for analytics execution.

5.4.1 Define Phase

As mentioned earlier, the Define phase focuses on project definition by establishing the scope, objectives, resources needed, and project plans with strong governance and stakeholder support. The project scope can be established by selecting the targeted business processes. In the define phase, the team must clarify the roles related to the project and identify all responsible individuals. The most important activity in the define phase is the creation of the project charter by establishing the scope, objectives, resources, expected business value, and role clarity. The project managers need to prepare a detailed project plan for the five phases of DMAIC, including tasks and associated deliverables. An example of a project charter and its elements is provided in Table 5.2.

TABLE 5.2
Project Charter Elements

Charter Elementt	Description	Comments		
Project Goal	Clear goal statement with a success mesaure. Goal should be Specific, Measurable, Attainable, Realistic and Time based (SMART goal)			
Scope	Project scope should include processes, tools, systems, and data that will be included. Primary metric:			
Resources/Roles	Esttimates of resources and associated roles.			
		Phase	Major deliverables	Date
Timeline	Project major deliverables with various phases and timelines	Define		dd/mm
		Measure		dd/mm
		Analyze		dd/mm
		Improve		dd/mm
		Control		dd/mm

5.4.2 Measure Phase

In the Measure phase, the first step is to identify the metrics/KPIs and associated measurement systems. For any metric, we should consider the following three aspects (Figure 5.5):

1. **Target**: This indicates target performance level, which is the ultimate destination
2. **Capability**: Attainable performance level, after removing the sources of variation
3. **Actual**: Actual performance level in the presence of variation.

Target refers to the target level of performance. Target setting is to be done by business experts through historical data analysis and comparisons with best in class. Capability refers to your ability to reach a particular level without any variation. It represents the attainable level of performance. Actual refers to the level of performance in an existing situation (with variation). Please note that, most often, the actual performance level is less than capability because of the inevitable nature of variation. We cannot completely eliminate the effect of variation completely; we can only minimize it. As the famous quality expert W.E. Deming, aptly put: *"Variation is life and Life is variation."* (Source: Deming 1993).

Variation usually comes from two sources- special causes and common causes. Simply put, special cause variation describes unique external factors that may cause a shift in the process or system. Common cause variation describes a cause inherent to the system being measured.

To get accurate measures, performing measurement system analysis (MSA) is quite important. MSA is a method for evaluating how much variation is present in the process by which we collect data. It answers questions regarding the accuracy and reliability of data, as well as the process of data collection. MSA also helps in evaluating the effectiveness of the data collection plan.

After performing MSA on the metrics, the next step is to identify the critical variables that are impact the metrics. For these variables also, MSA and DQ checks should also be performed as needed.

A very important activity related to data quality is data quality assessment. This is an iterative effort involving processing the results, validating them with subject matter experts, and refining the data quality rules as needed to drive those results. This work takes input from those who understand the data sets and their users. Here, we need to answer questions like "is a completeness score of 92% acceptable?" or "is 90% accuracy good enough?'. Applying SPC concepts (Jugulum 2014) is a good way to determine thresholds. These statistically determined thresholds should be adjusted

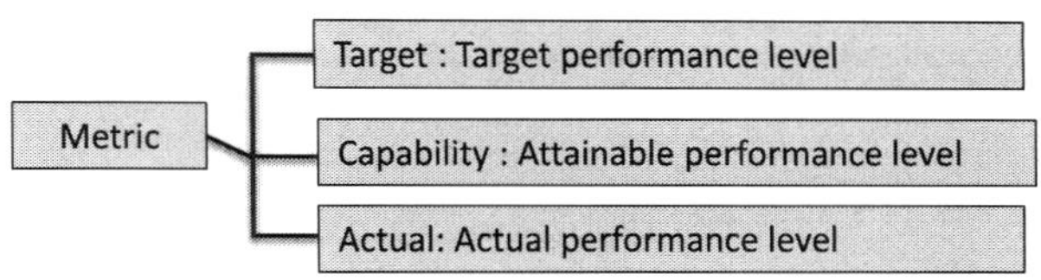

FIGURE 5.5 Measurement aspects of the strategic metrics.

as per SME inputs to obtain more practical thresholds to work with. Sometimes, when data is not available, we need to resort to SME inputs initially and refine the threshold as we collect more data. After determining dimensional level thresholds, a similar approach can be used for obtaining thresholds for overall DQ scores of CDEs.

The next step in the measure phase is to develop a conceptual analytical model with inputs and outputs. A parameter or p-diagram representation, as shown in Figure 5.6, is very useful here as well.

The p-diagram is very useful in clearly representing data sources, output from the model and variables that are controllable and uncontrollable, which will greatly help when fine-tuning the variable settings for better output. Understanding which variables to control will also help in capability and potential analysis.

5.4.3 Analyze Phase

After satisfying all data requirements of the measure phase, we then have to decide on the type of analytics that we need to perform for the conceptual model. Depending on the need, a suitable type of analytical model/technique needs to be selected, as shown in Table 5.1. There are several techniques/tools that can be used in the development of analytical models. These techniques include genetic algorithms, machine learning and fuzzy logic techniques, artificial neural networks, decision trees, cluster analysis, factor analysis, regression analysis, principal component analysis, discrimination and classification analysis, time series analysis, correlations and association analysis, test of additional information or Rao's test, and Taguchi methods.

5.4.4 Improve Phase

The improve phase of the DMAIC approach focuses on validating the models with actual outcomes and making adjustments to the model as needed. The outcomes should be practical and analytically sound. This phase also focuses on the creation of visual displays in the form of dashboards and scorecards. At the end of this phase, we should have a clear plan for operationalizing with appropriate steps.

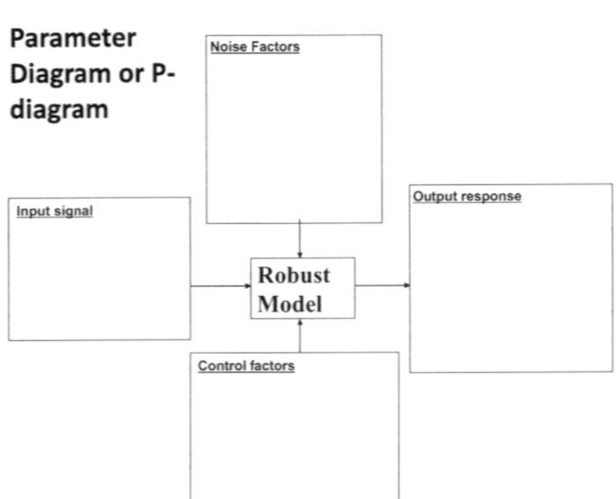

FIGURE 5.6 P-diagram representation.

5.4.5 The Control Phase

The control phase should focus on monitoring and controlling model outputs and corresponding impacting variables. The use statistical process control charts is highly recommended in this phase. Additionally, we should also periodically make sure that DQ levels are maintained and may need to re-run models with new information to check for model validity. This ensures that high=quality data and insights are maintained.

5.5 PURPOSEFUL ANALYTICS[2]

Any type of analytics should have a purpose, and its usage should be threefold. One aspect should focus on scientific discovery and advancement; the second aspect should focus on business decisions deployed through analytical insights; and the third and very important aspect should focus on societal needs by reducing loss to society. For any of these aspects, the creation and use of insights through analytics should be purposeful. The main purpose can be accomplished by identifying solution to the problem and acquiring knowledge for future research or improvement. The term "purposeful" is based on the following topology:

- Simplicity of usage
- Scalability
- Insights Accuracy and Precision

In this section, we discuss the importance and various aspects of purposeful analytics, why it is so relevant in the data-driven world, and how it helps achieve robust quality in analytics and in decision-making activities.

Figure 5.7 shows various characteristics of purposeful analytics. Their descriptions are provided below:

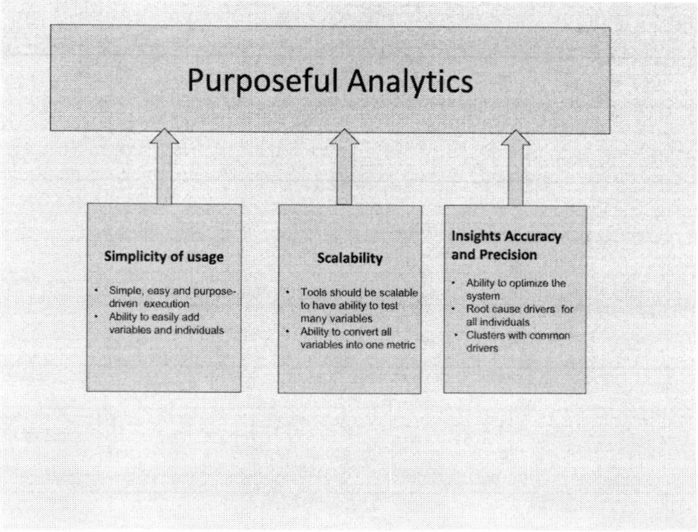

FIGURE 5.7 Characteristics of purposeful Analytics.

Simplicity of Usage: This refers to the importance of having simple and easy-to-run models and easy-to-interpret results. This aspect is extremely important as business decisions are made by business leaders who might not have exposure to advanced analytics or statistics. Simple and easy-to-run models must also have the ability to easily add variables and individuals

Scalability: Scalability is measured by the ability to test many variables at the same time with different forms of data numerical, text, voice, etc. and by the ability to convert all variables into a single metric. The tool that is being used should meet scalability requirements in performing analytical operations.

Insights Accuracy and Precision: It is extremely important to obtain insights precisely at the individual level by understanding key drivers affecting the performance. To determine these get key drivers, the analytical tool should have the ability to optimize the system by reducing the number of variables. Based on root cause drivers, one should be able to create clusters with common drivers so that suitable programs can be developed.

5.5.1 Individualized Analytics versus Population Based Analytics

The insights derived through analytics should be as individualized or personalized as much as possible to understand the behavior aspects at the individual level rather than at group or population level. Most analytical approaches cannot provide individualized insights as they are usually based on population analytics. Individualized analysis will help in making suitable decisions based on individual requirements.

Personalized/individualized insights are very important for purposeful analytics, as they are targeted at each individual separately rather than considering all individuals together as a population or group. Individualized solutions help organizations take suitable actions to enhance the satisfaction at the individual level. If the number of individuals is high, then we can identify the variables or combinations of variables causing the problem or abnormal behavior can be identified, and corrective actions can be planned accordingly. Table 5.3 summarizes some of key differences between individualized analytics and population-based analytics.

TABLE 5.3
Individualized Analytics versus Population Based Analytics

Individual Analytics	Population Based Analytics
• Not classification problem • Most often the goal is to see how far individuals are from target state • Root cause drivers for individuals separately • Groups (clusters) based on individual contributions for developing programs	• Classification problem • Same coefficients and equation is used for everyone • Root cause drivers for the group or population of individuals • We start with groups (clusters) and put individuals in them for developing programs

5.5.2 Examples of Individualized Insights

A medical facility decided to improve the process of diagnosing various health issues to proactively identify and resolve problems at an early stage through a better health management system. A medical specialist led this effort with a small team, which can be considered as the data science team. When it comes to measurement, the main purpose is to measure the overall health of the patients and determine necessary actions. Since the target metric should accurately measure overall health by developing a measurement system, capability and potential of the metric are not applicable.

Several variables including age, height, weight, total cholesterol, LDL cholesterol, HDL cholesterol, Triglycerides, Blood pressure and the number of hours of sleep were considered. The data corresponding to these variables was collected and ensured to be of good quality. Using this data corresponding to various individuals, a measurement system was developed. Here measurements were taken from a group of individuals who did not have health issues (desired state).

The data science team built a measurement scale based on the data from selected variables. The reference for this scale is derived from all of these variables for

TABLE 5.4
Root Cause Analysis with Contribution Ratios (Sample Data)

Notation	Variable	individual 1	Individual 2	individual 3	individual 4	individual 5
	Scale values	60.5	44.2	45.1	38.0	49.2
X1	Total cholesterol	1.87%	0.60%	4.95%	13.07%	1.13%
X2	LDL cholesterol	3.75%	2.85%	1.72%	0.24%	0.02%
X3	HDL cholesterol	0.36%	1.30%	0.42%	4.53%	9.77%
X4	Triglycerides	7.63%	0.95%	5.67%	4.25%	2.20%
X5	Non-HDL-C	10.47%	6.69%	21.36%	9.08%	4.25%
X6	ApoB	2.72%	6.26%	2.61%	0.14%	8.26%
X7	sdLDL-C	0.65%	1.21%	3.65%	1.21%	1.94%
X8	%sdLDL-C	2.87%	0.01%	8.95%	3.84%	4.45%
X9	VLDL-C	0.70%	7.49%	1.01%	3.95%	0.44%
X10	Lp(a)	1.44%	6.68%	0.24%	0.01%	2.85%
X11	ApoA-1	0.14%	29.30%	0.00%	0.78%	0.46%
X12	High Sensitivity CRP	0.33%	0.04%	0.63%	13.90%	7.32%
X13	ESR	0.00%	7.05%	0.63%	0.78%	0.00%
X14	HbA1c	0.04%	6.85%	2.55%	0.01%	5.22%
X15	glucose	11.43%	0.23%	0.90%	0.04%	7.08%
X16	insulin	12.11%	0.31%	0.53%	10.29%	2.60%
X17	mercury	0.14%	0.07%	0.00%	0.08%	0.09%
X18	Height	2.21%	3.30%	2.46%	1.77%	0.21%
X19	Weight	6.75%	0.58%	1.80%	0.29%	5.27%
X20	BMI	5.31%	0.69%	14.25%	2.25%	9.16%
X21	Waist Circumference	11.73%	0.79%	1.05%	9.51%	1.78%
X22	Most recent Blood Pressure Reading-Upper	0.61%	7.34%	5.89%	11.66%	0.03%
X23	Most recent Blood Pressure Reading-Lower	11.85%	2.52%	1.22%	0.61%	18.46%
X24	Minutes of exercise / week	1.02%	1.67%	0.19%	1.28%	0.37%
X25	Hours of sleep / night	2.10%	2.69%	15.33%	5.04%	5.13%
X26	cadmium	1.77%	2.53%	1.97%	1.41%	1.51%

healthy patients. Here, higher scale values indicate a higher degree of abnormality or unhealthiness. Table 5.4, shows the drivers corresponding to a sample set of patients with health issues. Note that these individuals also have higher scale values, indicating that they are abnormal or have health issues. Table 5.4 also provides details of variation analysis and contribution percentages of variables impacting the scale. In this table, values with contributions above 5% are highlighted.

5.6 MEASURING ANALYTICS QUALITY

As mentioned earlier, the improve phase of the DMAIC based data science approach focuses on validating the models with actual outcomes and adjust the model as needed. Usually, the models are validated based on the following metrics:

We usually use a confusion matrix or truth tables to calculate model validation measures. Combining the two actual positives/negatives and predicted positives/negatives will help in the creation of a confusion matrix. The structure of the confusion matrix is as shown in Table 5.5. Here, TP and FP correspond to true positives and false positives, while TN and FN correspond to true negatives and false negatives.

Below are some of the measures that we usually calculate using confusion matrix.

Recall is the ratio of positives that are correctly evaluated by the model to all actual positives in the data. – TP/(TP + FN)

Accuracy is the ratio of correct predictions provided by the model to all predictions – (TP + TN)/(TP + FP + FN + TN)

Precision is he ratio of correct positive predictions provided by the model to all positive predictions – TP/(TP + FP)

Let us say for a model, we have the following confusion matrix (Table 5.6). For this example, the different model validation measures are as follows:

Total sample size = 50

So, TP + FN + FP + TN = 50

$$TP = 20;\ TN = 25;\ FP = 3;\ FN = 2$$

TABLE 5.5
Confusion Matrix

PREDICTED VALUE		ACTUAL VALUE: Positives	ACTUAL VALUE: Negatives
	Positives	*TP*	*FP*
	Negatives	FN	TN

TABLE 5.6
Confusion Matrix Example

PREDICTED VALUE \ ACTUAL VALUE	Positives	Negatives
Positives	*TP* = 20	*FP* = 3
Negatives	*FN* = 2	*TN* = 25

$$\textbf{Accuracy} = (\textbf{TP} + \textbf{TN}) / (\textbf{TP} + \textbf{FN} + \textbf{FP} + \textbf{FN}) = \textbf{0.9}$$

$$\textbf{Precision} : \textbf{TP} / (\textbf{TP} + \textbf{FP}) = \textbf{0.87}$$

$$\textbf{Recall} : \textbf{TP} / (\textbf{TP} + \textbf{FN}) = \textbf{0.91}$$

5.7 ENHANCEMENT OF MODEL ACCURACY THROUGH DQ MEASUREMENT

In this section, we will discuss enhancement model accuracy measurement by using the robust quality index (RQI), which takes into account both analytics quality and data quality aspects. Consider three CDEs/Factors that are required inputs for an analytical model. For these three factors, we need to examine DQ levels by defining suitable DQ dimensions. Table 5.7 shows the DQ dimensions along with the corresponding scores and overall DQ scores.

After obtaining DQ scores for the metrics, in the next step is to examine analytics quality (model accuracy) from the model outputs. For measuring RQI for analytics execution, it is required to calculate deviations of DQ scores from the target and analytics output from the actual values. Since DQ scores are measured on a scale of 0 to 100, analytical outputs and targets should also be transformed to the scale of 0 to 100. The measure accuracy is calculated as percentage, making it easier to transform it on a scale of 0 to 100. After doing this and defining targets, deviations will be calculated and from the deviations signal to noise (S/N) ratios are calculated. Based on S/N rations, analytics RQIs are measured on a scale of 0 to 100 with score 100 indicating the best model. Table 5.8 shows details of analytics RQI calculations for one model. For this model, accuracy or model output is 75%

TABLE 5.7
DQ evaluation of three CDEs/Factors (Scores are in Percentage)

DQ dimension	Completeness	Conformity	Validity	Accuracy	DQscore
CDE1/Factor 1	96	93	88	87	91
CDE2/Factor 2	95	90	85	78	87
CDE3/Factor 3	92	88	84	76	85

TABLE 5.8
Analytics RQI for a Model

CDE/Factor	Scores	Target	Squared deviation
CDE1/Factor 1-DQ	91	95	16
CDE2/Factor 2-DQ	87	90	9
CDE3/Factor 3-DQ	85	90	25
Model output	75	100	625
		MSD	13.0
		SNR (log)	-11.1
		RQI for Analytics	44.3

If multiple models are used, the overall analytics quality can be determined by aggregating the RQI results of all models. Table 5.9 illustrates the process of obtaining analytics quality in a two-model scenario. Details of RQI calculations are given in Appendix-I. Readers are encouraged to contact the authors for any questions regarding RQI calculations.

TABLE 5.9
Overall Analytics RQI

	CDE	Scores	Target	Squared deviation
Model 1	CDE1/Factor 1-DQ	91	95	16
	CDE2/Factor 2-DQ	87	90	9
	CDE3/Factor 3-DQ	85	90	25
	Model output	75	100	625
Model 2	CDE1/Factor 1-DQ	86	90	16
	CDE2/Factor 2-DQ	93	96	9
	Model output	70	100	900
			MSD	15.12
			SNR (log)	-11.80
			RQI for Analytics	41.02

NOTES

1 Thanks are due to CRC Press for allowing us to use materials from prior publications of one of the authors.

2 The authors (Jugulum & Heien) are thankful to dataDragon/Lumenytics team since they worked with them while developing some aspects of this section. The author thank Wiley for granting permission.

REFERENCES

Davenport, Thomas H. and Harris, Jeanne G 2007. Competing on Analytics- The New Science of Winning, HBS press.

Deming, W. E. 1993. *The New Economics: For Industry, Government and Education.* MIT press.

Gartner (2012). Big Data Strategy Components: IT Essentials. Gartner publication. Published on 15th October 2012.

Jugulum, R. 2014. *Competing with High Quality Data: Concepts, Tools and Techniques for Building a Successful Approach to Data Quality*. Wiley publication.

Jugulum, Rajesh 2018. Robust Quality: Powerful Integration of Data Science and Process Engineering, CRC publication

Phadke, Madhav S. 1989. Quality Engineering Using Robust Design, Prentice Hall, New Jersey, Englewood Cliffs.

6 Big Data and Cloud Solutions

6.1 HISTORY OF BIG DATA

As children, some of the first adjectives and antonyms we learn in English are big/small, fast/slow, high/low, and good/bad. The first of these words, 'big', represents the paradigm shift in how we conceptualize the size of data that society is able to capture as technology continues to advance. Big data has always existed in the world, expressed in our everyday lives and interactions, and the ability to capture it efficiently has only accelerated through the turn of the century. As there will always be 'more' data, the future expanse of its capture may be driven less by 'can we' and more by 'should we' given the rise in data breaches, carbon dioxide production through energy consumption, accumulation of e-waste, and the societal impact of the products of Big Data (e.g. Generative AI, personal data collection, mass surveillance.)

6.1.1 Timeline of Data Collection

The often cited first use of data for analytics is the collection of bubonic plague data in 1663 by John Graunt, cited as the father of Demography, and author of *'Natural and Political Observations Made upon the Bills of Mortality'* which documented his observations of published mortality rates titled 'Bills of Mortality' printed on paper.

As technology advanced, data, as well as instructions (the precursor to modern day 'software'), could be encoded in the form of punch cards, invented by Herman Hollerith in 1884.

The 20th century brought a rapid acceleration in data storage mediums. In 1928, Fritz Pfleumer invented the magnetic tape drive, followed by the programmable computer Colossus in 1943, which was used to help decipher encrypted Nazi messages. In 1965, a United States committee of the Social Science Research Council developed The National Data Center Proposal to promote sharing of large, structured data sets within the federal government. Four years later, the United States developed the Advanced Research Projects Agency Network (ARPANET), the first wide-area network, which established protocols still used in the modern Internet.

With the creation of the fundamental building blocks of distributed data processing (data storage, processing, and computer networks), the concept of Big Data began to emerge in the final decade of the 20th century: the Founding of the World Wide Web, the maturation of cost-effective digital storage, Google is founded and rise to in dominance in the search industry, and the introduction of NoSQL. The century concluded with the titular question posed in Peter Lyman and Hal R. Varian's book

DOI: 10.1201/9781003190325-6

"How Much Information," which sought to quantify the amount of digital data stored in the world: 1.5 Billion Gigabytes (Lyman & Varian 2000).

6.1.2 Relation to Software

In terms of computing, the general assumption is that 'Big' is always 'Good' when it comes to data, based on the very nature of the invention of computers: to compute, or process, more and faster to save humans time in revealing knowledge and insights. In this chapter, we will assume that 'Low' cost, 'Fast' processing, and 'Big' data are 'Good' aligning with the metrics of the Project Management triangle's measure of Cost, Time, and Quality in developing a project, being in this case a Cloud software application (Atkinson 1999).

In the following diagram, any project will make tradeoffs between these three measures. Below are four types of projects based on their proximity to the relevant measure. The scenarios will provide example of consumer products rather than projects to help conceptualize where these trade-offs might occur in our everyday lives:

A: Moderate Cost and Speed to Produce, but lacking in quality. This product was created with constrained resources and has a high tolerance for defects and consumer dissatisfaction. This can be exemplified by a screwdriver made for nonprofessional use in the home. It may be easily mass-produced and only needed on a few occasions by the customer. To improve quality, many manufacturers may produce a higher quality product as a baseline, then identify semi-defective yet functional units for sale at a lower quality, a process known as 'binning' parts.

B: Moderate Cost and Quality, but takes a long time to produce: Seasonal fruits and vegetables are an example of a product that requires high to moderate quality, for health reasons, moderate cost due to farming labor, yet is always constrained by the time needed for growth. In this instance, speed of production might be increased through various solutions, such as off-season storage and preservation.

C: Moderate speed to product and quality, but high Cost (low affordability.) This focus on quality and speed with a high budget is best exemplified by high cost, high -risk projects such as airplane manufacturing. Quality is a non-negotiable to ensure the safety of passengers during flight, while speed of production is critical to remain competitive in the airline booking business. Cost containment measures can be implemented through microservices associated with the primary service, such as additional purchase charges for luggage, in-flight entertainment, and fees for rebooking and flight cancellations. In addition, some incentives may further reduce costs, such as profit sharing from in-airport services and income tax deductions from the city's services to the airline.

D: This product represents a balance of all measures formerly discussed. Each scenario above mentions a mitigation strategy to help balance its respective shortcomings to bring product feasibility into balance (Figure 6.1).

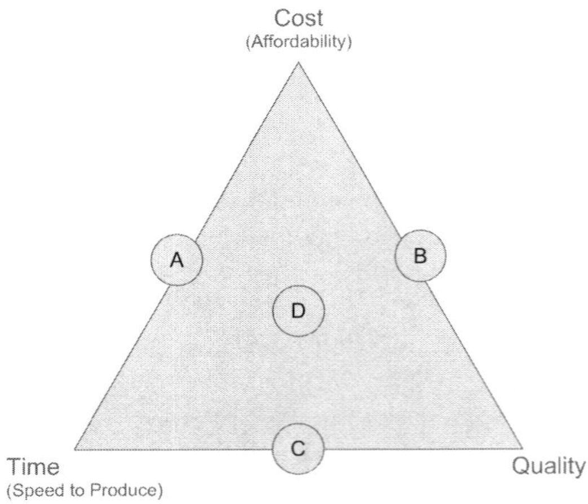

FIGURE 6.1 Project management triangle.

6.2 TECHNICAL COMPONENTS OF DATA CAPTURE

As mentioned in the history of Big Data, data is observed, recorded, and processed in order to be of use. These represent the technical steps in which data is utilized, and have the following properties that must be considered when designing any digital project:

1. **Sensing**: The energy expressed by a real-world object or process that produces a signal of data.

   Example: A digital camera contains a rectangular, flat silicon sensor, constructed of thousands of photosites, areas that can absorb a photon of light. When taking a picture, each photosite records either a 1 (photon detected) or 0 (no photon detected), effectively recording light as a grid of bits (0 or 1). The camera's processor collects all the bits from the photosites and sends them to the camera's storage as an image (Figure 6.2).
2. **Signal Encoding**: The agreement on the interpretation of signals to record from the real world observations. For digital systems, the signals produced from sensing are a sequence of binary values: on or off, 1 or 0. This is required both when writing (recording) data and reading data back data from a storage medium.

   Example: The bits recorded from a digital camera's sensor are organized by the camera's processor, which append additional data, known as metadata, around the recorded bits to identify the collection as a standard image format and of a certain width and height (Figure 6.3).
3. **Storage Medium**: The physical medium to which an encoded signal is written. This length of time the data is maintained, or persisted, within the medium may vary based on the quality of the medium used. Typically, higher quality mediums have longer lifespan, greater durability, and higher cost.

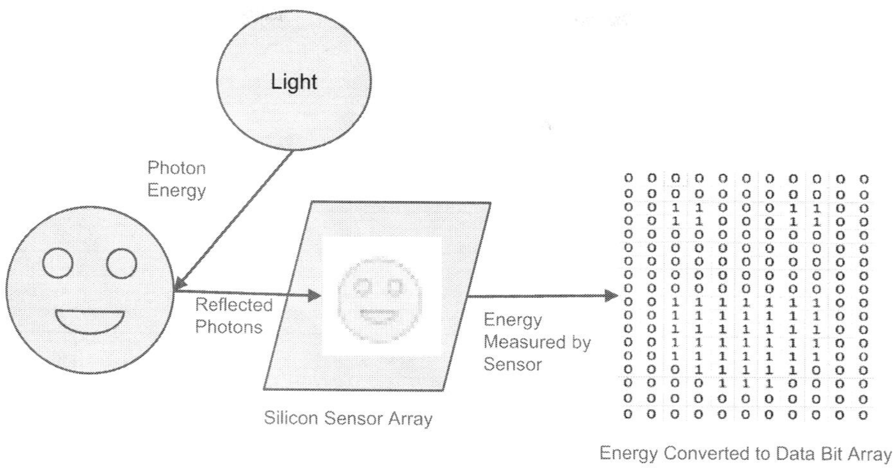

FIGURE 6.2 Sensing an image.

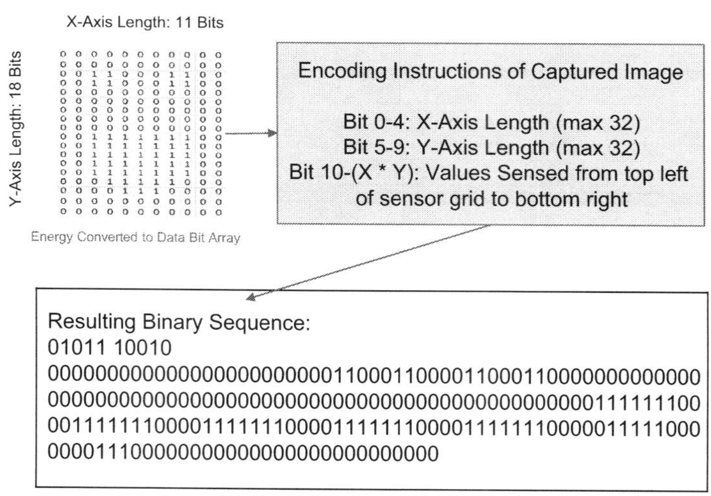

FIGURE 6.3 Encoding sensed bits based on encoding instructions.

Example: A basic SRAM (Static Random Access Memory) memory card is a silicon chip, housed in a memory card casing, and used in a digital camera to record the encoded signal and metadata into a 'file', which remains stored even when the camera is powered off.

4. **Storage Format**: The specific method a storage medium uses to prepare its raw material to accept digital data. Most storage formats lay out physical boundaries (clusters) to which data (bits) can be written, while maintaining a record of all bits currently written. This ensures that existing data is not unintentionally overwritten, and data to be read can be correctly located.

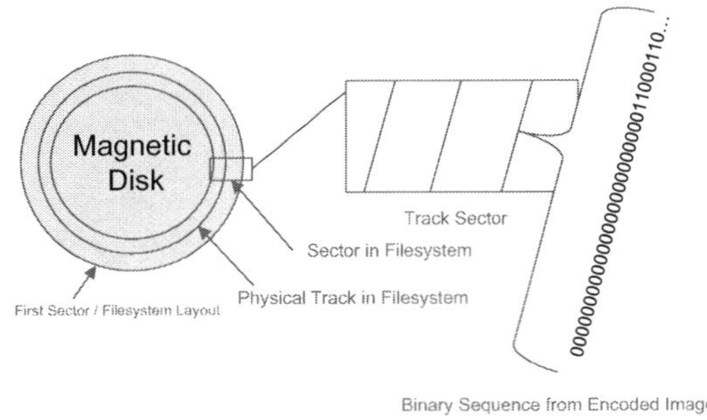

FIGURE 6.4 Storage medium & image data location.

Example: The memory card used in the digital camera is formatted using the File Allocation Table 32 (FAT32) filesystem specification. The digital camera's processor contains software that is able to read and write from FAT32 formatted medium. The camera's processor reads that there is free space to write at cluster location 5400–5652. The image file size will take 200 clusters to store (cluster count and size may vary depending on formatting options and storage medium size), and is written to cluster location 5400–5600. Cluster index 5400 is used to begin writing the file's metadata, including the 'header' information detailing what type of data is stored at that location, and how many clusters to read (200) in order to interpret the entire file (Figure 6.4).

5. **File Format**: Whereas a medium's Storage Format defines how files are organized, read, and written on the physical medium, a file format defines how bits from the encoded signal are arranged within a file and interpreted by applications. Many file formats exist for the signal data of a digital photo, and offer different degrees of quality and size depending on how the bits from the original signal are enhanced. Additional processing on raw image signals may allow for reducing the file size (i.e. compression), increasing image size and quality through algorithms (e.g. interpolation), and storing a various degrees of complexity (appending of geolocation data, color-space information, and time/date of photo), depending on if the application reading the file can support the file format.

   Example: A digital camera allows storing in multiple formats based on a user's intent to load and view files in other computer systems that support reading the common format. RAW format is one example of a camera-specific format, capturing the entirety of sensor data to be recorded for further refinement in software applications, yet it takes several seconds to write to the storage medium. This format is preferred for professional, low-speed scenes such as portrait or still-life photography, as the quality and level of detail are more valuable than the time needed between photos taken. JPEG format (a standard developed by the Joint Photographic Experts Group) results in a smaller file due to compression and loss of detail (i.e., a lossy format) of the original sensor image, yet allows many photos to be taken quickly. This format may be preferred for high-speed events and photo images intended for use in digital publications.

6. **Retrieval**: The process by which a system and application retrieve and interpret data of a specific format from a formatted storage medium and reproduce the signal as originally recorded.

   Example: A digital camera typically had a built- in Liquid Crystal Display (LCD) screen for the user to view the photos previously taken. The processor within the camera used to record, format, and store the file may also be the same processor used to view the photos after they have been recorded. The only difference is that the LCD is now displaying a reproduction of what was recorded. This entire process replicates the light, or image, a human eye might see at a point in time.
7. **Volatility**: The length of time data persists within a given medium. Different components of a system may be built either as volatile (data is lost when electricity is no longer applied) or non-volatile (data persists even if not receiving electricity.) Volatile storage is usually limited to memory chips within computers and only meant for temporary storage.

   Example: A digital camera may have a 'preview' mode, allowing the user to preview an image before choosing to save it to the memory card. The camera's processor only needs to perform the Sensing and Signal Encoding steps outlined above, storing the image in the processor's on-chip memory before showing the image to the user. Since this reduces the number of steps involved, previewing an image is much faster than physically storing and reviewing the image. However, the image will be unrecoverable if the camera is powered off before saving the previewed image.

6.2.1 Data Capture Scenarios and Big Data Storage

In the previous examples, each stage of data capture and storage was described through the process of capturing a digital image. However, many businesses collect data in less concrete scenarios yet still retain the common elements of data capture and storage. Below are two additional examples of the seven components of data capture and storage aligned with that of digital image capture: A scenario involving door-to-door census of a town to create population density and cohabitation demographics, and the recording of a medical patient's height and weight as part of a routine physical at a physician's office (Table 6.1).

6.3 HISTORY OF CLOUD COMPUTING

The term 'cloud' in 'cloud computing' most prominently represents the concept of a fully abstracted level of computing service, one where the hardware, software, and processing of data are distributed, 'always on' due to built-in redundancies, scalable to the workload needed, requiring network connectivity, and is offered at varying levels of security. Each of these aspects alone does not define cloud computing; rather the combination of these factors, along with advances in speed and the lowering costs of commodity technology, has contributed to its rise as a distinct concept.

TABLE 6.1
Various Scenarios Incorporating the Components of Data Sensing and Storage

Component	Digital Camera	National Census	BMI Measures for Medical Patients
Sensing	Silicon Sensor	Record of Resident Details (Age, Names, Relationships, Address) in Ledger	Patient weight on Scale and height when standing against a ruler
Signal Encoding	Camera Processor recording Energy	Voice of resident responding to questionnaire heard by census taker and transcribed to text	Nurse visibly inspecting scale and ruler and transcribing to keyboard input on computer
Storage Medium	Storage Disk (Solid State Silicon Medium)	Pen and Paper	Computer Database (Hard Disk Drive: Rotating Magnetic Medium)
Storage Format	Encoded File	Entry of the Primary Resident per line, up to 6 sub-record of co-habitants, up to 8 households per page	Weight with two decimal points of precision, height in inches with one decimal point of precision
File Format	JPEG Format	Sequential Entry in Ledger	'INSERT' query into Database Management System (DBMS)
Volatility	Preview: High - lost on power loss Storage: Low - up to 50 years on magnetic medium	Acid-free Paper: Up to 100 years in controlled climate	Storage: Low - up to 50 years on magnetic medium

6.4 TECHNICAL COMPONENTS OF CLOUD COMPUTING

6.4.1 Distribution

It is important to note that distributed systems (interconnected computers sharing data for parallel storage processing) have existed long before the idea of cloud computing. Distributed systems were necessary due to the limited resources of a single computer (or node) within a network. Just as a single builder may take several years to fully construct a house, one commodity node (low cost, non-specialized hardware) within a network is insufficient to store and process large amounts of data in a timely fashion. When combined into a network, nodes share the data and workload required without investment in a single, high cost system.

6.4.2 Redundancy

The combination of nodes may or may not include fault-tolerance in the form of redundancy. Nodes may share copies of the same data, and work on the same inputs in parallel to mitigate the loss of data being processed in the event of node failure. Redundancy also involves a cost investment in duplicate hardware and software for the given nodes.

Three areas in which the need for redundancy is typically assessed are storage (persistence of data on a medium such as a hard disk), and processing (applying logic and transformations on data once read from a storage medium.)

6.4.2.1 Storage Redundancy

A single node can enable redundancy of storage for various purposes depending on the use of the data. The following examples are ordered by the recovery time required from the redundant copy of the data:

- **Archiving**: Data is archived with no expectation that it will be required in the near future. Archiving is often required by policy to enable future audits or retrieval for legal proceedings.

  Example: The Internal Revenue Service (IRS) within the United States has varying requirements for archiving tax data by type for individuals and corporations. [https://www.irs.gov/businesses/small-businesses-self-employed/recordkeeping]
- **Back-Ups**: Data with potential use in the near future, in the case of hardware failure, theft, or corruption of data, can be copied in its entirety to a separate location as a backup. The concept of a backup may take several forms, as long as the process of making a point in time replication of data. A single file may be backed-up, or a disk, or an entire node across all disks. Another distinction of a backup is that it is a batch process, meaning the job of backing up is done once to completion and may be scheduled for a full, or incremental, run over a period of time.

  Example: Microsoft Windows operating systems can be scheduled to take routine backups (daily, weekly, monthly, etc.) of the operating system in case of system corruption due to the installation of a virus, malware, or other system-corrupting software.
- **Real-Time Storage Replication**: Expanding on the concept of a backup, methods exist to replicate redundant copies of data in real-time. Examples of this method include RAID Mirroring, short for Redundant Array of Inexpensive (or Independent) Disks, in which a microcontroller directs data to multiple physical disks in parallel. In the most basic example, a twin pair of disks has single redundancy, meaning that if one disk fails, the system can continue to operate with no data loss although redundancy is lost until the failed disk is replaced.

6.4.2.2 Processing Redundancy

In addition to storage redundancy, processing redundancy focuses on the ability for in-process data to be replicated and preserved in the case of a full node failure. Similar to storage redundancy, processing redundancy has several implementations, ranging from high-time-to-recovery to low-time-to-recovery.

- **Virtual Machine Clusters**: A Virtual Machine (VM) involves a host system that uses software to virtualize one to many hardware systems, each with its own localized and virtualized CPU, RAM, and storage. These VMs operate on shared resources of the host system, they may be paused, backed-up, archived, or replicated with little effort. Many identical VM nodes may be clustered to achieve node-level redundancy as a cluster, with data being processed on the virtual cluster instead of the underlying host. Although total processing capability is slower than if performed on the host due to the overhead of the virtualization software, the in-process data is preserved in the event of a node failure.

  Example: Oracle's VirtualBox and ProxMox (based on Linux's Kernel Virtual Machine) are examples of software virtualization systems able to replicate identical nodes along with virtualized networks in which to establish clusters.
- **High-Availability (HA) Clusters**: High-availability refers to the concept of an end user of any system experiencing little or no interruption to their interactions with said system. In High Availability clusters, two or more identical physical, networked nodes are built to each hold and process an identical set of data and software. Using physical nodes, as opposed to virtual nodes, adds additional redundancy when considering downtime: Physical power outages, hardware faults, or networking outages. Technically, a virtualized High Availability cluster achieves the same objective; however, the difference from Virtual Machine clusters is that High Availability. The assembling of multiple identical nodes, along with software systems to manage the network and failover mechanisms between these nodes, refers to High Availability systems. High Availability is required for cases that demand little or no interruption to services for users.

  Examples: High traffic public-facing websites, flight-control systems, mainframes, and stock trading platforms.

6.4.3 Scalability

Modern advances in hardware components, such as power-saving features, variable clock speeds (or variable power consumption), and other dynamic energy efficiencies allow our modern computers to save power when idle, or not in use. These efficiency features, when combined with the concepts of Virtual Machines, result in a system that can automatically shut down unneeded virtual nodes as workloads decrease, and, conversely, add additional nodes as workloads increase. The underlying hardware power costs of the VM host. These components of a cloud allow cloud providers to offer a scalable system with a predictable price, creating a viable market for Cloud providers to offer services to organizations and individuals.

6.4.3.1 Economics of Scale

A traditional Data Center includes a facility real estate costs, utility costs, power costs, computer hardware costs, and staffing costs. These costs are incurred even if the owning company's utilization of the data center fluctuates. In contrast, Cloud services are offered to many organizations and can adjust prices to match costs on demand.

> Cloud computing involves nodes that are expected to be connected to some form of network. A public cloud, as opposed to a Data Center or private cloud, indicates that all underlying infrastructure costs are owned directly by the user and incurred upfront at a fixed rate, rather than billed at a variable rate based on usage.

A comparison can be made to show trade-offs in these two approaches. The first illustrates the larger fixed cost when there is low utilization relative to resource cost, while the second table shows how scaled up utilization can quickly increase the billed cost of public clouds (Table 6.2)

TABLE 6.2
Public Cloud vs. On-Premises Cost with Partial Utilization for Eight Hours

Resource	Utilization per Day	Private Cloud Cost	Public Cloud Cost	Public Cloud Cost
CPU	8 hours	$500	$0.25/hr	$2.00
Memory	8 hours	$3,000	$0.75/hr	$6.00
Data Transfers In	1,000 GB transfers	$0	$0.01/GB Transferred	$10.00
Data Transfers Out	1,000 GB transfers	$0	$0.01/GB Transferred	$10.00
Data Storage	10,000 GB Storage	$200	$0.001/GB Storage	$10.00
Facility Costs	36 KWh @ $0.20 per KWh	$7.20	$0.00	$0.00

It becomes evident that Public Clouds can quickly incur higher costs due to their on-demand pricing, however there are several advantages beyond hardware and facility costs that add to a public cloud's value (Table 6.3).

TABLE 6.3
Public Cloud vs. On-Premises Cost with Full Utilization for One Year

Resource	Utilization per Year	Private Cloud Cost	Public Cloud Cost	Public Cloud Cost
CPU	8,760 hours	$500	$0.25/hr	$2,190.00
Memory	8,760 hours	$3,000	$0.75/hr	$6,570.00
Data Transfers In	365,000 GB transfers	$0	$0.01/GB Transferred	$3,650.00
Data Transfers Out	365,000 GB transfers	$0	$0.01/GB Transferred	$3,650.00
Data Storage	10,000 GB Storage	$200	$0.001/GB Storage	$10.00
Facility Costs	13,140 KWh @ $0.20 per KWh	$2,628.00	$0.00	$0.00
Total Cost per Utilization for 1 year		$6,328	$16,070.00	

6.4.4 Public, Private, and Hybrid Clouds

The key differentiating factor between private and public clouds is the owner and maintainer of the physical infrastructure. As shown in the prior comparison of costs, private cloud infrastructure and operation are wholly owned by the user, whereas public cloud infrastructure is a shared architecture in which any number of users may access resources based on a pre-agreed billing fee schedule. These two approaches cater to various states in which a cloud approach offers value to a business or individual.

6.4.4.1 Public Clouds

A public cloud is a service-based approach to providing all the previously mentioned technical components of cloud computing, abstracted away from the end user's view. No physical hardware must be directly managed, maintained, or purchased by end users. Rather, customers only manage the arrangement of these resources to suit their specific application. Many of these users' resources may reside on the same physical hardware and be given either a percentage of computing power (reserved pricing) or a time-specified pricing model based on demand when compute resources are needed (on-demand pricing.) (Amazon AWS 2024).

In some applications, businesses may build services on top of existing public cloud infrastructure. For example, Apple leveraged AWS resources, fully owned by Amazon but licensed by Apple, to host iCloud cloud-based backups for Apple customers (Thompsons Data 2024). Many businesses offer various cloud-based services via the public cloud, whether for cloud storage, such as Google Drive, DropBox, Apple iCloud, or Microsoft OneDrive, or processing services, such as music streaming through Spotify or Apple iTunes.

6.4.4.2 Private Clouds

In contrast to publicly available cloud resources purchased for a fee, private cloud infrastructure is maintained by the entity using these resources. This responsibility includes infrastructure such as power, cooling, and facility security, as well as hardware specific costs such as repairs, maintenance, redundancy, and financial accounting of assets including depreciation costs. Many of the complexities in hosting a private cloud come with the benefit of having full and total control over the amount of compute power used and distributed among the users of the private cloud. There is an added benefit of data privacy, as many public cloud infrastructures are subject to security breaches and risk leaking customer data in unsecured cloud resources (Goodin 2024). A private cloud may choose to limit access to internal networks or VPNS, as a method to mitigate hacking attempts by reducing the 'Internet Facing' threat mentioned in IBM's article describing various digital attack surface vulnerabilities (IBM 2024).

Private cloud architecture and scale vary widely depending on the scope of users, resources, and access type needed. The various components of private clouds offer flexibility in terms of implementation, with the technical components scaled to the need. Below are a few examples sorted by expected demand, from low demand single-user setups to enterprise-grade accessibility by organizations of 500+ employees:

1. **Individual or Consumer-Grade**: One of the most basic types of private clouds is a singular device, such as a consumer-grade range of NAS (Network Attached Storage) devices consisting of 2 or more hard disk drives connected and accessible via the Internet (QNAP 2024). This device has some degree of redundancy and can be accessed via the public Internet through various methods, such as a vendor-provided portal or a private-network-to-public VPN (Virtual Private Network.)
2. **Small-Office/Home-Office**: Larger scale, multi-user private clouds involve pairing dedicated servers with fault-tolerant or fail-over routing equipment and firewalls to ensure accessibility (Allin 2019). The level of redundancy increases in accordance with the expected load and wear on equipment, in addition to increased network bandwidth, such as load-balancing, to accommodate multiple connections from users.
3. **Data Center**: The largest implementation of a private cloud consists of many servers located in an owner-maintained facility with advanced access controls implemented, such as identity management and intrusion detection. This may be considered a private cloud if users can access resources anytime, from anywhere, maintaining the always-available aspect of 'cloud' resources.

6.4.4.3 Hybrid Clouds

The prior sections differentiated public from private clouds; however, these resources can be intertwined as needed based on a product's demand for cost, time, and quality, as outlined at the beginning of the chapter. The following examples detail various scenarios in which public and private resources are combined to achieve the most cost effective means of delivering a cloud application:

1. **Private Development, Public Deployment**: Many startups may lack seed funding when first developing a cloud-native application. To minimize costs, they can invest in lightweight private cloud infrastructure during the development and debugging stages of creating a minimum viable product, or MVP, prior to further outside investment. This fixed-cost approach allows multiple development cycles and, if developed using a common cloud packaging method such as Kubernetes, can be seamlessly deployed and scaled in the public cloud for production (Chase 2021).
2. **Backing and Disaster Recovery**: While a private cloud offers securely and privately, there may be a need for secure, low-cost effective offsite replication for backup or disaster recovery. To maintain security, the private cloud owner can encrypt all offsite backup with private keys, guaranteeing the public cloud data storage provider has no access to view underlying data and removing the risk of a data leak from the public cloud provider (Mnev 2021).
3. **Software Escrow Services**: When a cloud product is developed as a private cloud application, investors may require access to the underlying code and infrastructure if the startup fails to meet performance goals. In such cases, a public cloud provider's software escrow services can act as a neutral third party to securely host code and application instances, ensuring continuity and preservation of assets (Kesslen 2020).

6.4.4.4 Cloud Storage, Applications, and Analytics

The emergence of Software-as-a-Service business models have further enabled new companies to build on top of Cloud capability with applications over a vast array of distribution models. Applications built on cloud platforms can deliver a wide range of always-on, always-available services such as entertainment, document storage, photo sharing, and social media interactions to name a few.

If a product is expected to be widely accessible and always available on-demand, then it is most likely one form of cloud architecture upon which that product is delivered. With many applications using commodity processing hardware, cloud providers must differentiate themselves either through price or capability. One aspect of capability emerging is the race to have the fastest and most capable processors offered for artificial intelligence or quantum computing applications. These resources represent unique capabilities that cloud architectures, both public and private, can offer users at a premium (Chernicoff 2023).

With ever-increasing competition among cloud providers, continuous advancements are being made in how new and innovative applications and products are distributed. As cloud computing resources become more accessible, more capable, and cost-effective for anyone with Internet access, the barriers of cost, time, and quality can be increasingly easy to overcome, lowering barriers to entry for anyone and everyone to develop new cloud-based solutions.

REFERENCES

Allin, D. October 2019. Dual WAN Policy Based Routing with a USG. https://help.gowifi.co.nz/support/solutions/articles/48000960320-dual-wan-policy-based-routing-with-a-usg. Accessed September 29, 2024.

Amazon AWS. What's the Difference Between On-Demand Instances and Reserved Instances? https://aws.amazon.com/compare/the-difference-between-on-demand-instances-and-reserved-instances. Accessed September 29, 2024.

Atkinson, R. December 1999. Project management: cost, time and quality, two best guesses and a phenomenon, it's time to accept other success criteria. *International Journal of Project Management*. 17 (6): 337–342. doi:10.1016/S0263-7863(98)00069-6.

Chase, N. 2021. The New Stack "Private vs. Public Cloud: How Kubernetes Shifts the Balance". https://thenewstack.io/private-vs-public-cloud-how-kubernetes-shifts-the-balance. Accessed September 29, 2024.

Chernicoff, D. 2023. Data Center Frontier. "How Intel, AMD and Nvidia are Approaching the AI Arms Race". https://www.datacenterfrontier.com/machine-learning/article/33002883/how-intel-amd-and-nvidia-are-approaching-the-ai-arms-race. Accessed September 29, 2024.

Goodin, D. July 2024. Ars Technica. "Hackers steal 'significant volume' of data from hundreds of Snowflake customers". https://arstechnica.com/information-technology/2024/06/hackers-steal-significant-volume-of-data-from-hundreds-of-snowflake-customers. Accessed September 29, 2024.

IBM. What is an attack surface? https://www.ibm.com/topics/attack-surface. Accessed September 29, 2024.

Kesslen, M. P. 2020. Law360 "Source Code Escrow Agreements Are Reaching For The Cloud". https://www.lowenstein.com/news-insights/publications/articles/source-code-escrow-agreements-are-reaching-for-the-cloud-kesslen-satlin. Accessed September 29, 2024.

Lyman, P. and Varian, H. R. 2000. How much information? Technical Report, School of Information Management, UC Berkeley. Reprinted in Wirtschafts Politische Blatter, 48, Jahrgang 2001.

Mnev, A. 2021. AWS Security Blog "Demystifying KMS keys operations, bring your own key (BYOK), custom key store, and ciphertext portability". https://aws.amazon.com/blogs/security/demystifying-kms-keys-operations-bring-your-own-key-byok-custom-key-store-and-ciphertext-portability. Accessed September 29, 2024.

QNAP. 2024. "myQNAPcloud: The best way to remotely access your NAS. Simple & Secure." https://www.qnap.com/en-us/software/myqnapcloud. Accessed September 29, 2024.

Duff, J. 2023. Thompson Data. "Top Fortune 500 Companies that Use AWS". https://www.thomsondata.com/blog/fortune-500-companies-that-use-aws/. Accessed September 29, 2024.

7 Structuring Unstructured Data and NoSQL

7.1 RECAP OF STRUCTURED DATA

Data must typically be structured in some form, such as a table within a database management system (DBMS), before being used for any reporting or analysis. These tables contain data in a well-defined structure of rows, also referred to as records or observations of the subject of the analysis, and columns, also referred to as attributes or features of each observation recorded from the subject.

Additional details of the types and formats of data recorded in each table, referred to in a database management system as a table's schema, will not be discussed in depth, instead this chapter will focus on how and why data is recorded in structured and unstructured formats.

In its simplest form, the records contained in a table are observations of some real-world entity or interaction recorded line by line. The features of the subject recorded are often defined prior to observing the real world data, an aspect of the standard Scientific Method approach to research.

The key aspect of this structured format is knowing the set of attributes you wish to collect prior to data collection and, in turn, knowing that these attributes can be recorded for each entity observed. To exemplify the need for structured data before collection, we can consider a simple use case of improving residential apartment building security for residents.

7.1.1 Sample Use Case: Apartment Building Security Survey

An apartment owner may receive complaints of crime within the building and wish to invest in better security features for the building, such as alarms, locks, and security cameras. The owner may receive only a few specific complaints of break-ins but may not know any other details of these events other than the residents' names and descriptions of the incidents. The owner decides to conduct a building-wide census by door-to-door survey.

To make decisions on the types of security most needed to improve security for residents, the owner will need to know what questions to ask each resident through a survey form. In other words, he must 'structure' his questions before recording responses. Asking the question 'From whom or what will I receive my answers?' will identify individual records represented in the structured data table. Asking, 'What answers do I need to complete my analysis?' will form the columns in which data is recorded.

Without a survey structure, residents may respond in any number of 'unstructured' ways: lengthy stories about their life in the apartment unit, their opinions of other

DOI: 10.1201/9781003190325-7

TABLE 7.1
Simple Survey Form and Results Stored in a Data Table

Name	Age	Apartment Unit	Gender	Theft in Past 6 Months	Assault in Past 6 Months	Noise Complaint in Past 6 Months
John	32	1A	M	No	No	No
Sarah	27	3B	F	No	No	Yes
Bob	64	2A	M	No	No	No
Cindy	44	2B	F	Yes	No	No
Sam	89	3A	M	No	No	No
Emily	48	1B	F	No	No	Yes

neighbors, unnecessary demographic data (height, weight, hair color), problems they have with other amenities such as plumbing and heating, and countless other possibilities of data unrelated to criminal activity they have observed. By limiting the data recorded and knowing what data will be uniformly asked across all residents, the owner can complete the survey in a quick and efficient manner (Table 7.1).

Each question or field on a survey form represents an attribute of the structured data of the resident in relation to their individual experience with building security. Each resident's response is recorded within a row of the data. An alternate design for this survey may involve recording multiple instances or types of crime within the building that the resident has observed, but for simplicity, the owner limits resident responses to one or none.

Once the survey design is completed, the apartment building owner has effectively created what is known as a data model for one table to capture the data needed to answer his question. To reiterate, the emphasis of this activity is that a data model must be created prior to the collection of data; otherwise, the owner may receive any number of useful responses that are not relevant to building security.

Looking at the data, we can immediately identify some relevant and irrelevant data regarding the safety of the apartment building. No residents have experienced assault in the past six months, so this attribute may be effectively removed from further analysis of this dataset. Only one theft has occurred, in addition to several noise complaints. The building owner can now identify and categorize these occurrences as high risk, in the case of theft, and low risk, in the case of noise, to resident safety. Before the owner makes assumptions about how to prevent future theft and noise issues, there may be deeper meaning in this data if combined with unstructured and semi-structured data. The following sections will describe additional unstructured and semi-structured data that the owner can distill from various sources to support the best solution.

7.2 OVERVIEW OF UNSTRUCTURED AND SEMI-STRUCTURED DATA

Now that we have a fresh understanding of the simplest form of structured data, the table, we know that it is created at the time of need for analysis, includes assumptions about the data needed for analysis being necessary and available from the data source, and is limited to only the data defined in the structure. However, with the rise of Big Data, we have access to datasets and observations from a wide variety of sources provided by governments, research institutions, vendors, corporations, and individuals. The ability to process this large quantity of structured data easily and affordably, along with digitization efforts of paper-based records, has opened the potential to explore additional forms of data for analysis beyond the purpose for which it was originally captured: semi-structured and unstructured data.

7.2.1 Unstructured Data

Observations of our world exist all around us and are recorded in various methods, regardless of any analytical purpose or business decision needed by humans. The rings of a tree are interpreted as a record of its age (Helmenstine, 2024), the magnetic polarity of volcanic rocks records the geomagnetic reversal of Earth's magnetic poles (Lahr, 2016), and at one point or another, we, as humans, may have captured our own daily experiences in a descriptive log such as a diary or journal.

The characteristic feature of unstructured data is that there is no preconception of the form, format, or length of attributes expressed by a given entity—only that the observations are recorded in some form, either intentionally or unintentionally.

To illustrate the capture of unstructured data, we can add one additional column to the original survey from Table 7.2, named 'Notes,' which is used to capture any miscellaneous comments in a variable-length text format from apartment residents. The fact that this data can be of any length and structure allows flexibility in the survey to store information that may or may not be useful in the final analysis. However, it is assumed that extra effort will be needed to parse out any discovered information captured in this open-ended input.

7.2.1.1 Feature Extraction from Unstructured Data

In reading through entries in the 'Notes' field, the apartment owner may find keywords or phrases relevant to the issue of crime in the building. Note that this survey only has six total respondents, so it is a low-cost activity to manually read through each note given by residents. In this case, we will explore one method of extracting information from this unstructured text using pattern matching, known as regular expressions.

TABLE 7.2
'Notes' Field Added to Capture Free-Form, Unstructured Text

Name	Age	Apartment Unit	...	Theft in Past 6 Months	Noise Complaint in Past 6 Months	Notes
John	32	1A		No	No	Recently started a new job at...
Sara	27	3B		No	Yes	Complained about noise and music...
Bob	64	2A		No	No	Worried rent will increase next...
Cindy	44	2B	...	Yes	No	Returned home at noon on Wed...
Sam	89	3A		No	No	Retired and usually visits the park...
Emily	48	1B		No	Yes	Heard loud banging noises upstair...

7.2.1.1.1 Regular Expressions

A regular expression is a set of typed character notations evaluated by a regular expression engine to match a target piece of text or a substring of text within a larger body of text (Crutchley, 2022). Regular expression usage for feature extraction exemplifies a rule-based approach to identifying desired subtext. That is, the creator of the regular expression explicitly defines the subtext patterns they are searching for rather than relying on a data-driven approach such as machine learning (Wisbey, 2024).

In the following excerpts from the 'Notes' field, the apartment owner is attempting to extract any relevant information related to noise and thefts within the building, as those types of complaints were reported as experienced by residents:

- Cindy, theft reported: "Returned home at 12pm noon on a Wednesday, August 14th, to find her apartment door had been forced open. Upon entry and inspection, she found her locked bedside nightstand had also been forced open with pieces of jewelry missing."
- Sarah, noise reported: "Complained about noise and music coming through the window from outside at 11:30pm on a Tuesday."
- Alice, noise reported: "Heard loud banging noises upstairs around 11am on Wednesday morning."

The above text contains several potentially useful pieces of data that, if extracted, could be added to our dataset and used as additional support for a solution to prevent future crimes. Each note mentions some form of time that is relevant to the complaint of the resident. Since the owner recorded the times in a consistent numerical format, regular expressions can be used to search for and extract these times. For this, the regular expression must accommodate the following assumptions:

1. The hour of time is stated as numeric integer
2. The hour is not written using 24 hour format, rather uses the am/pm suffix
3. A minute measure may or may not be added to the hour, delimited by a colon character

For these three assumptions, we can use the following regular expression (Jagannath 2023):

```
\b([1-9]|1[012]|)(:[0-5][0-9])?(\b)?(i?)(am|pm)\b
```

To summarize this sequence, the prefix and suffix '\b' represent word boundaries, which should enable this pattern to search on a word-per-word basis within the note. Next, the leftmost parenthetical grouping searches for one of the numbers in the range 1–9, or 10, 11, 12. For the second grouping, a colon, the first digit of '0–5', and the second digit of '0–9' are searched for, collectively tagged as optional by the '?' character. The third grouping represents an optional word boundary to allow a space to precede the 'am' or 'pm' suffix. The fourth grouping switches the statement to case-insensitive evaluation, effectively converting the fifth grouping to a search for all case variations of 'am' and 'pm': 'AM', 'PM', 'aM', 'Am', 'pM', 'Pm' (Table 7.3). Most computing languages contain functions to implement regular expression

TABLE 7.3
Resulting Data Extracted from Descriptive Text

Notes	Time of Event
Returned home at 12 pm noon on a Wednesday, August 14th, to find her apartment door had been forced open. Upon entry and inspection, she found her locked bedside nightstand had also been forced open with pieces of jewelry missing	12:00 pm
Complained about noise and music coming through the window from outside at 11:30 pm on a Tuesday	11:30 pm
Heard loud banging noises upstairs around 11 am on Wednesday morning	11 am

evaluation. Table 7.3 shows the expected results if this regular expression pattern were fed into any of these functions, with the returned result stored in a new column of data called 'Time of Event'.

It is important to note that regular expressions are constructed using assumptions about the structure of patterns sought. In this regard, they may be useful for extracting certain explicit pieces of information for a very narrow purpose. In the above instance, the time mentioned is assumed to be in relation to noise or theft experienced by residents. Residents may mention other time-based information that may not be applicable, yet it is still detected by the regular expression defined. Residents may not have described theft or noise, yet the regular expression is still evaluated (Table 7.4).

TABLE 7.4
Unintentionally Extracted Data from Descriptive Text

Notes	Time of Event
Recently started a new job after moving to the city 2 months ago.	
Worried rent will increase next year and asks if he can lock in a lower rate if the owner is available to meet after 3 pm.	3 pm
Retired and usually visits the park until sunset around 8 pm.	8 pm

This process has not extracted data that is incorrect based on the assumptions made. However, this can be remedied by filtering the data to include only those with 'Yes' responses for residents with a theft, assault, or noise complaint. Another complication may arise if a resident has multiple complaint types, and the time extraction method from notes has not been affiliated with any specific event type. These are some limiting factors in the use of regular expressions, yet they exemplify one method to create structured data from unstructured text.

If this dataset were expanded to larger buildings, neighborhoods, or entire cities, the resulting thousands or millions of records would require the implementation of more sophisticated techniques to accurately extract desired features as new columns of data, a process known as feature extraction (Paolo 2019).

7.2.1.2 Natural Language Processing

Another type of structuring method for text descriptions is natural language processing (NLP). This approach relies on underlying dictionaries of grammar and syntax to drive the interpretation of parts of speech within a body of text, allowing for more precise extraction of structured data. Many modern tools are available to extract various aspects of free-form text, such as Amazon Comprehend (How to analyze insights in text with Amazon Comprehend | AWS 2022).

This, and other NLP solutions like it, can detect several types of structured output based on a large underlying corpus of knowledge and may use both rules-based and statistics-based approaches to determine meaning (Natural Language Processing (NLP) [A Complete Guide] 2023).

Using our original text entries from the above example, we can explore additional points of data useful for the analysis. Although the exact algorithms used to evaluate NLP-extracted data won't be covered in this section, below are examples of the types of entity information that can be extracted. Note that each of these extractions of information is based on parts of speech such as nouns and adjectives only within the given row of text (Table 7.5).

Each row, or observation, is evaluated independently from one another, with no context-awareness between observations. For example, the 'Tuesday' information extracted may, in truth, be August 13th, the day prior to the first complaint of theft. The 'Wednesday' mentioned in the third record may be the same day as the first, August 14th, but without setting additional context and assumptions, such as in the case of machine learning systems, these connections between records cannot be made. In the next section, we explore adding relationships to each individual record in order to form more context around these events, and to determine if those relationships add any value to our analysis.

7.2.2 Semi-Structured Data

Semi-structured data, as the name suggests, has a form that is less strictly defined in tabular format than the table structure previously discussed, yet more structured than

TABLE 7.5
Additional Structured Data Extracted Using NLP

Notes	Time of Event	Source of Event	Evidence
Returned home at 12 pm noon on a Wednesday, August 14th, to find her apartment door had been forced open. Upon entry and inspection, she found her locked bedside nightstand had also been forced open with pieces of jewelry missing	12:00 pm Wednesday, August 14th	Apartment and Nightstand	Forced Open
Complained about noise and music coming through the window from outside at 11:30 pm on a Tuesday	11:30 pm Tuesday	Window	Music
Heard loud banging noises upstairs around 11 am on Wednesday morning	11:00 am Wednesday	Upstairs	Loud Banging

open descriptive text. The implementation of semi-structured data storage allows a balance between performance and flexibility, with the opportunity to add additional structure based on the data needed.

7.2.2.1 Log Data

One example of semi-structured data is system log data, or the capture of events over time that occur in a given scenario. The structure of the underlying log store may be generic enough to capture multiple event types in a less complex schema. Some NoSQL, or unstructured DBMSs, are MongoDB and CouchDB (Tuguran 2024). For example, motion detection systems may have the log format shown in Table 7.6, which captures multiple event types in a simplified structure.

This log alone is useful in tracking the observance of the camera; however, it lacks information about what events transpired around it in relation to the motion detection. As it stands, the purpose of this camera is to act as an indicator to the observer to investigate further if needed. Modern security cameras are affordable enough to install within doorbells or other residential features in many homes.

The sophistication and affordability of consumer-grade cameras are ever increasing, to the point where the actual observed images (unstructured data) can be analyzed with image recognition technology, with resulting tags appended to logs to better inform the observer. Tags are simple words or phrases that have been manually associated with various images as observed by a human. These tags collectively agree on what content exists in each image and form a basis for the statistical probability of a given image containing a certain element or object. Several large collections of tagged images exist to facilitate image recognition ("About ImageNet 2021"; "Amazon Rekognition Image" n.d.).

Adding these additional tags of information to the schema above is trivial, and due to the semi-structured nature of the storage schema, requires no modification of the underlying database or table columns. Below is a recreation of the above camera log example, but with enhanced 'Activity Type' descriptions based on tags returned from an image recognition system, based on the images captured by the camera.

With the above semi-structured data, we can now explore how we might integrate such log data into our continued analysis of apartment crime. Let's assume the apartment owner had one such camera capturing data similar to Table 7.7 to produce this log for the 16 hours leading up to the theft reported at 12:00 p.m. on August 13th.

TABLE 7.6
Example Log of a Motion Detection Camera

Activity Time	Duration	Activity Type
12:30:26.135	5s	Motion Detected
12:30:23.663	20s	Motion Detected
12:12:06.567	1s	System Online
12:11:46.349	20s	System Reboot
12:11:45.163	5s	Motion Detected

TABLE 7.7
Example Log of a Motion Detection Camera with Image Recognition Feature

Activity Time	Duration	Activity Type
12:30:26.135	5s	Package Delivery Detected
12:30:23.663	20s	Animal Detected
12:12:06.567	1s	System Online
12:11:46.349	20s	System Reboot
12:11:45.163	5s	Person Detected

Upon reviewing this log data in isolation, the apartment owner may note that there is a large assumed outflux of identified persons passing by the camera early Wednesday morning, with many also detected, assumed to be influx, throughout the hours of 8 to 9 p.m. the prior evening (Table 7.8).

To reiterate, the motion detection feature and activity type tagging are meant as an indicator to the owner to further investigate if needed. In this case, there is a great need for any data, including this semi-structured log data, as this activity can shed light on the potential root cause of theft and noise reported. In the next section, we'll combine all types of data we've reviewed up to this point to reveal the exact conditions in which a crime took place. This will serve to inform the owner of the next best investment to make for building security should these conditions occur again.

7.2.2.2 Graph Data

One lingering question the owner may have is why half of the residents of the apartment experience no noise complaints, as sound is expected to have ambient traversal throughout a building. One semi-structured form of data, not directly stored within a table structure or able to be queried using standing Structured Query Language (SQL), is a graph.

Graph databases represent entities depicted as vertices and relationships between these entities depicted as edges. A graph enables exploration of relationships, with queries 'traversing' the graph to return those relationships that best

TABLE 7.8
Log Data from the Apartment Entrance Camera

Date	Time	Activity Type
Wednesday, August 14th	12:00 pm	1 Person(s) Detected
Wednesday, August 14th	11:05 pm	1 Person(s) Detected
Wednesday, August 14th	2:00 am	16 Person(s) Detected
Tuesday, August 13th	9:00 pm	4 Person(s) Detected
Tuesday, August 13th	8:45 pm	7 Person(s) Detected
Tuesday, August 13th	8:15 pm	2 Person(s) Detected
Tuesday, August 13th	8:05 pm	3 Person(s) Detected
Tuesday, August 13th	8:00 pm	1 Person(s) Detected

match the query submitted. One such simple query is, "Who are the friends of my friends?" This basic question is one best suited for graph structures. If this question were to be asked of a relational database, it would require a full join of entities against themselves, a compute-heavy operation that is costly for table-based DBMSs. Some examples of graph databases are Cassandra, Elastic Search, and Neo4j (Kovacs 2010).

Graphs may be created with any type of relationship as a focus, and many graphs may be constructed using the same relational entities within a table. The base requirement of constructing a graph is the identification and assignment of entities as nodes and the definition of their relationships and all properties of both. Let's consider some graphs that can be constructed using the data we've collected from residents so far.

Figure 7.1 shows a very simple relation of each resident to the apartment building as a whole. Note that there is not much information contained in this graph aside from the concept of a resident 'living in' an apartment building. This graph also lacks any entity properties or additional information about each resident such as age or apartment unit. This graph alone does not answer any question aside from "Who lives in the apartment building?"

Next, we can consider different relationships between residents through use of the 'Age' attribute for each resident entity (Figure 7.2).

Although it's possible to create this graph from our survey data, the graph doesn't answer any questions relevant to the presence or root cause of criminal activity within the apartment building. This is a fundamental challenge of graph data: How should graphs be constructed from our structured data in order to answer a question, and provide additional calculated findings back to our structured data?

We can now revisit the apartment owner's original question, "Why did half of the residents not cite noise complaints on the night of Tuesday, August 12th, or the morning of Wednesday, August 13th?" Since the focus of this question is on residents and noise, we can construct our initial graph with the following relationship of resident to noise in Figure 7.3.

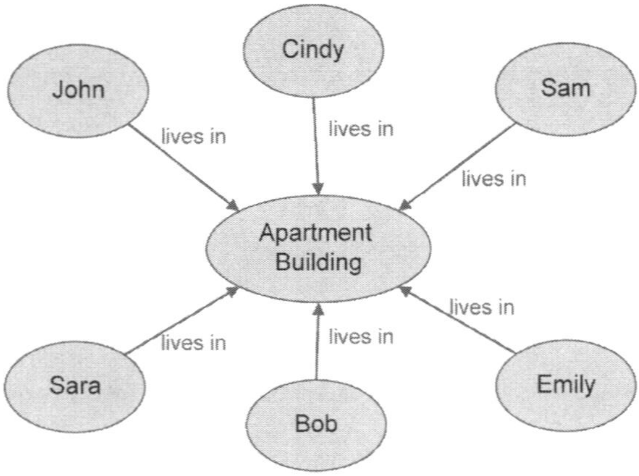

FIGURE 7.1 Graph structure showing the relationship of resident to the apartment building.

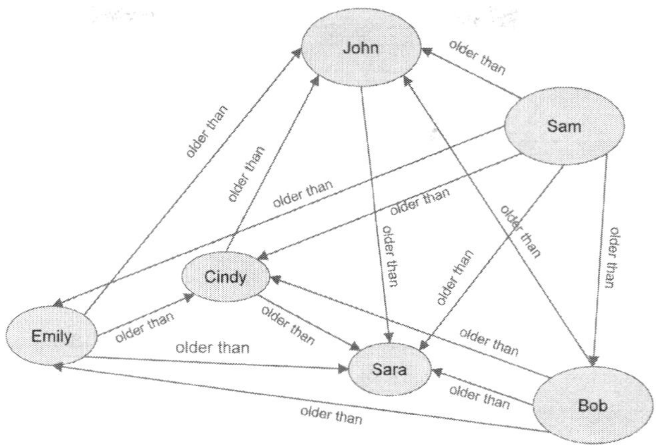

FIGURE 7.2 Graph structure showing the relationship of ages between residents.

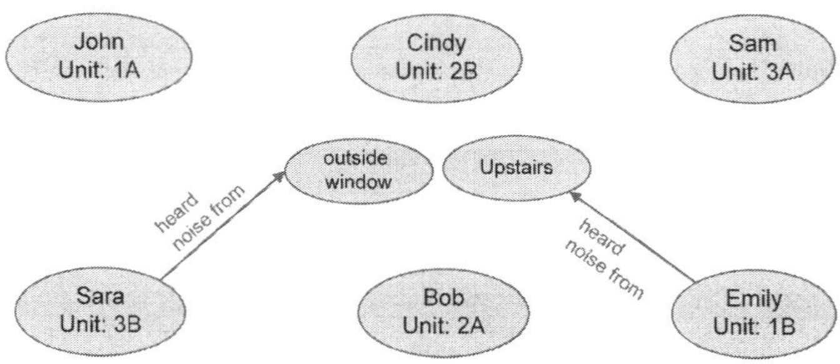

FIGURE 7.3 Graph showing the relationship of resident to the source of noise experienced.

Based on this relation, four of the residents will not have any relation defined, as no noise was experienced. Another important observation is that the nodes representing 'noise origin' have no connection to any other entity on the graph, which prevents this form of the graph from answering any questions.

The apartment owner considers his original question again, regarding sound, and considers the real-world entities that could relate sound from a resident to the activity within the building. The owner then realizes that the missing information for this graph is the physical barriers to sound throughout the building: walls, doors, floors, and windows.

To create a fully connected graph with this additional data, the owner must source one final dataset: the architecture of the building, detailing the placement and sound transmission levels of various materials throughout the building. In a related application, this type of data is also considered by renter insurance providers in terms of 'Firewall' materials separating units, with inflammable materials, such as brick, being considered lower risk for fire spread between units (Romero 2023). The owner is able to retrieve the building construction diagram illustrated in Figure 7.4.

This information can be added to the graph from Figure 7.3 to fully connect the graph in order to answer the owner's original question. In Figure 7.4, the materials

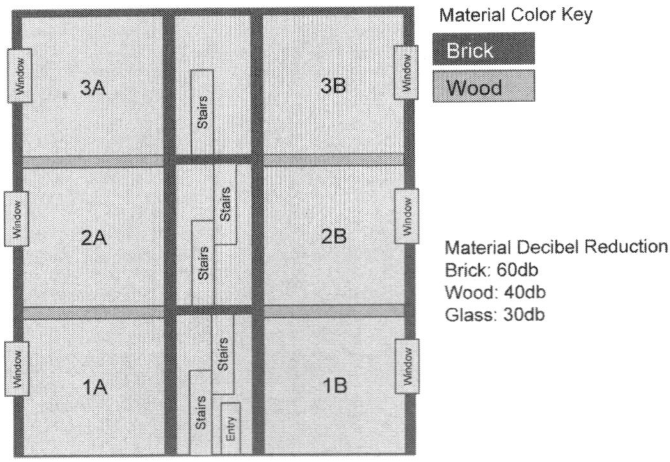

FIGURE 7.4 Architectural layout of building and materials separating apartment units.

are added in relation to the apartment unit positions, with the material properties of 'decibel reduction' added to the vertices. We'll assume for simplicity that any verticy not shown is inaudible between apartment units and materials, such as window pairs on opposite sides of the building.

With the graph in Figure 7.5, we are now able to fully traverse all apartment units on the graph and, more importantly, calculate a 'cost' metric for graph traversal. Graph traversal is a method of querying the graph based on the conditions of nodes and cumulative properties of their respective vertices. Many algorithms for graph traversal exist (Needham and Hodler 2019, Chapter 4); however, our question pertains to the distance, or proximity, of one resident to a source of noise. Therefore, we can use a shortest-path approach to ask the graph, "If a noise occurs at a given decibel level in one apartment unit, in which apartment units could that noise be audible and at what decibel level?" Conversely, this gives us the ability to also ask the question

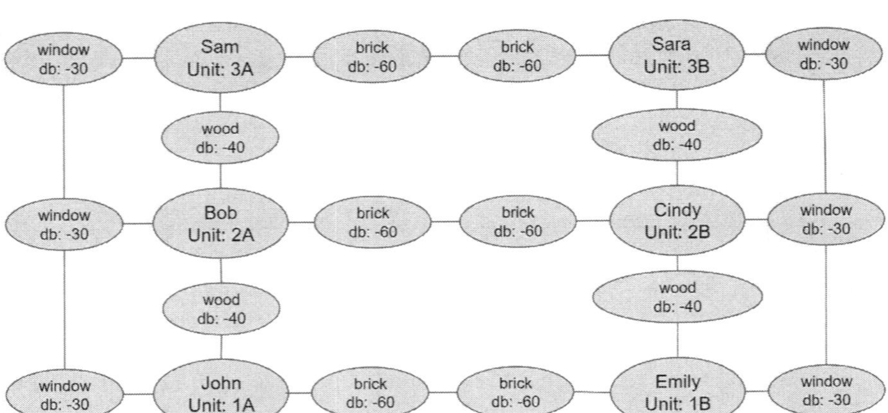

FIGURE 7.5 Graph will all decibel reductions between apartment units based on materials.

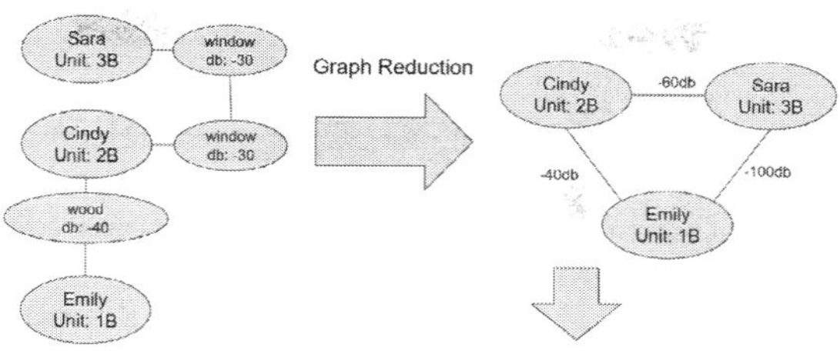

Final Structured Dataset

Source Unit	Target Unit	Traversal Cost
3B	2B	-60db
3B	1B	-100db
2B	1B	-40db
2B	3B	-60db
1B	3B	-100db
1B	2B	-40db

FIGURE 7.6 Total traversal cost of decibels across apartment units with noise complaints.

the apartment owner is seeking to answer: "If a noise is heard in multiple apartment units, what is the closest common unit from which that noise could originate?"

Since we know the two units that observed noise, we can reduce the graph to only those nodes that are relevant, along with the pathway through which those noises were observed, as shown in Figure 7.3.

The application of the graph calculations allows us to arrive at a final, structured form of data that we can now append back to our original structured dataset (Figure 7.6).

7.2.3 Combining Structured, Unstructured, and Semi-Structured Data

We can now assemble a fully structured table recording all data collected to this point:

1. Structured survey data responses from residents: We will exclude apartment numbers and records with no crime type reported. We will also remove the column for 'Assault' types of crime reported, as all respondents replied 'No'.
2. Structured times of noise and theft events occurring were extracted from unstructured text 'Notes' provided through resident interviews.
3. Structured volumes of persons in and out of the building between 8:00 p.m. on Tuesday, August 13th, and 12:00 p.m. on August 14th.
4. Structured features of relationships based on graph analysis were used to determine the minimum noise required to be heard across other apartment units in the building.

"The owner is now ready to assemble all information captured to this point in an effort to determine why noise and theft might have occurred and consider future remedies to mitigate future occurrences. Table 7.9 represents all structured data derived from unstructured and semi-structured sources, joined to the original structured questionnaire data after removing variables that were discovered to be irrelevant to the analysis: Name, Age, Gender, and Assault in the Past 6 Months. We can also

TABLE 7.9
Combining All Previously Unstructured Data to a Single, Structured Table

Apt Unit	Neighboring Unit	db Reduction between units	Theft Reported Time	Noise Reported Time	Db of Noise Heard	Neighbor Unit is Possible Origin of Noise?	Time Difference of Noise from Neighbor Theft	Total Persons Detected Entering in Timeframe	Total Persons Detected Entering in Timeframe
1B	2B	-40db	No	11:30pm Tuesday	50db	Yes	-12 hr 30 min	17	16
1B	3B	-100db	No	11:30pm Tuesday	50db	No	N/A	N/A	N/A
2B	1B	-40db	12:00pm Wednesd ay	No	N/A	N/A	N/A	N/A	N/A
2B	3B	-60db	12:00pm Wednesd ay	No	N/A	N/A	N/A	N/A	N/A
3B	1B	-100db	No	11:00am Wednesd ay	80db	No	N/A	N/A	N/A
3B	2B	-60db	No	11:00am Wednesd ay	80db	Yes	-1 hr	1	1

remove observations of residents who observed no criminal activity, as the graph analysis revealed that these apartment units were isolated in terms of noise observance from Tuesday, August 12th, to Wednesday, August 13th."

"Rows 1 and 6 are the most significant in the findings from this structured format. We can see that noise was observed late at night by apartment unit 3B the night before the theft, along with a large influx of persons detected on the front door camera. Most people were detected leaving by 2 a.m., except for one. The following morning, another noise disturbance was heard from the apartment downstairs of the apartment experiencing theft, with a person detected leaving prior to the resident of unit 2B arriving."

At this point, the apartment owner could attempt to hypothesize about the events occurring on the night of Tuesday, August 12th, such as apartment unit 2B hosting a party, as music was heard by the upstairs apartment along with a large influx of people earlier in the evening, and an outflux of people in the early morning, with one suspicious person remaining in the building overnight, possibly the thief. However, the intent of this exercise is to illustrate how one might go about structuring data in an attempt to observe and identify all critical and necessary variables surrounding the events in all their forms: structured surveys, unstructured notes from interviews, and semi-structured diagrams of the building layout and noise traversal across units. One prominent variable that could be considered when solving for future risks of this scenario is the influx of large numbers of people into the building. A mitigating factor for future risks of theft scenarios may be to require guest sign-in and sign-out processes, which in turn may require the hiring of personnel to enforce this added security measure at an added cost to residents' rental fees.

7.2.4 Intentional and Unintentional Data Capture

For unstructured and semi-structured data, the analytic use of the data may not be known at the time of capture, but it may have a need outside of a structured format. In addition, the data capture may never be used by the observer, or it may have unintended uses after capture. The following sections outline these scenarios and potential concerns regarding the data we unknowingly generate every day in our digital lives.

7.2.4.1 Intentional Data Capture

One final note on data in any of the forms previously mentioned involves intentional and unintentional capture. When an entity is to be observed, yet the availability and usefulness of the data are unknown, some form of intentional capture is defined. For example, a phone call between a customer service representative and one of the company's customers may record the audio from the support call to ensure the representative followed company policy to serve the customer to the best of their ability. The call may be reviewed by management in cases where the representative was reported to have a low satisfaction score from the customer.

In this use case, the call data has no immediate need unless a low satisfaction score is given by the customer. It may be stored in a digital audio recording and never replayed if all customers are satisfied. In the case where the call needs review, the audio is listened to and interpreted by a manager. There is no formal structuring of data taking place when listening to the audio, only the human interpreting evidence of undesirable behavior from the representative.

This data will remain captured as intended for its original use, yet it will not be structured if no further analysis is needed. It's important to now consider that many types of data, for purposes other than analytics and in these unstructured formats, are owned by companies looking to ever-increase their competitive advantage within their industry. As businesses exhaust novel ways to combine and extract value from structured data, intentionally captured unstructured data represents an asset already available but yet untapped for its potential until it is structured for a purpose.

7.2.4.2 Unintentional Data Capture, Metadata, and Exhaust Data

When an entity is being observed, yet data is captured unknowingly or not according to the needs of the observer, this data may have been recorded for a third party or simply included due to the low cost of capture. One common type of data recorded by digital sensors is metadata, or data describing aspects of the data being observed, not the recorded attributes the entity is expressing. Some examples of metadata are the conditions under which the data is captured, the time of capture, by whom or what the data was captured, etc. Often, this metadata is included for a purpose beyond that of the observer.

For example, many digital photos are captured in the modern age with a user's cellular phone, as the device contains both features. A cellular phone, by nature, contains location sensors to assist in cellular tower tracking and GPS (Global Positioning System) for map navigation. Taking photographs will commonly also record the location, time, and date from these additional sensors if the data is available. The user taking the photo may not be aware that this information is recorded, as their original intent is to observe a scene and record a picture. The phone manufacturer may realize the value of the metadata captured, and if multiple sets of it were to be structured alongside the user's personal demographic data, the manufacturer may have opportunities for more advanced analytics, such as detecting if the user is on vacation and suggesting nearby attractions or restaurants.

REFERENCES

About ImageNet. 11 March 2021. *ImageNet*. https://image-net.org/about.php. Accessed 29 September 2024.

Amazon Rekognition Image. n.d. *AWS*. https://aws.amazon.com/rekognition/image-features/. Accessed 29 September 2024.

Crutchley, C. 14 April 2022. The complete guide to regular expressions (regex). *CoderPad*. https://coderpad.io/blog/development/the-complete-guide-to-regular-expressions-regex/. Accessed 30 September 2024.

Helmenstine, A. 23 March 2024. Tree rings - dendrochronology and dendroclimatology. *Science Notes*. https://sciencenotes.org/tree-rings-dendrochronology-and-dendroclimatology/. Accessed 30 September 2024.

How to analyze insights in text with Amazon Comprehend | AWS. 16 May 2022. *AWS*. https://aws.amazon.com/getting-started/hands-on/analyze-sentiment-comprehend/. Accessed 29 September 2024.

Jagannath, S. 9 August 2023. Regex for time am/pm. *DebugPointer*. https://debugpointer.com/regex/regex-for-time-am-pm. Accessed 29 September 2024.

Kovacs, K. 25 December 2010. The big NoSQL databases comparison. *Kristof Kovacs*. https://kkovacs.eu/cassandra-vs-mongodb-vs-couchdb-vs-redis/. Accessed 29 September 2024.

Lahr, J. C. 7 December 2016. Sea-floor spreading and subduction model. *USGS Publications Warehouse*. https://pubs.usgs.gov/of/1999/ofr-99-0132/. Accessed 30 September 2024.

Natural Language Processing (NLP) [A Complete Guide]. 11 January 2023. *DeepLearning.AI*. https://www.deeplearning.ai/resources/natural-language-processing/. Accessed 29 September 2024.

Needham, M. and Hodler, A. E. 2019. *Graph Algorithms: Practical Examples in Apache Spark and Neo4j*. Beijing: O'Reilly Media.

Paolo, P. 10 October 2019. Feature extraction techniques. An end to end guide on how to reduce a… | by Pier Paolo Ippolito. *Towards Data Science*. https://towardsdatascience.com/feature-extraction-techniques-d619b56e31be. Accessed 30 September 2024.

Romero, M. 6 July 2023. Multi-family homeowners insurance (duplex, triplex, fourplex). *AgileRates*. https://www.agilerates.com/advice/home/your-guide-to-multi-family-homeowners-insurance/. Accessed 29 September 2024.

Tuguran, V. 8 January 2024. MongoDB vs. couchbase. *Baeldung*. https://www.baeldung.com/mongodb-vs-couchbase. Accessed 29 September 2024.

Wisbey, O. 12 June 2024. Choosing between a rule-based vs. machine learning system. *TechTarget*. https://www.techtarget.com/searchenterpriseai/feature/How-to-choose-between-a-rules-based-vs-machine-learning-system. Accessed 30 September 2024.

8 Design and Analysis of Multivariate Diagnostic Systems[1]

8.1 INTRODUCTION

We have earlier discussed the importance of ensuring high-quality data. After ensuring its quality, the data should be used in a meaningful way to derive important insights. In this chapter, we describe the Mahalanobis-Taguchi Strategy (MTS) and its applicability to developing a multivariate diagnostic system, as we deal with deriving information from multiple variables. Examples of multivariate systems include medical diagnosis systems, client relationship mechanisms, fraud detection systems, fire alarm sensor systems, and so on. The Mahalanobis distance (MD) is used to measure the distances in a multivariate system, and Taguchi's principles are used to measure the accuracy of the system and to identify important variables that are sufficient for the measurement system.

8.2 DESCRIPTION OF MAHALANOBIS-TAGUCHI STRATEGY

Mahalanobis-Taguchi Strategy (MTS) is a multivariate diagnostic-based analytical technique, which is used to conduct a system diagnosis through a multivariate measurement scale. The Mahalanobis distance takes into account the correlation between the variables. In multivariate systems, patterns are extremely sensitive to correlations between the variables, so the use of MD is quite appropriate to capture the correlation information. The importance of correlations in multivariate systems can be explained by using Figure 8.1

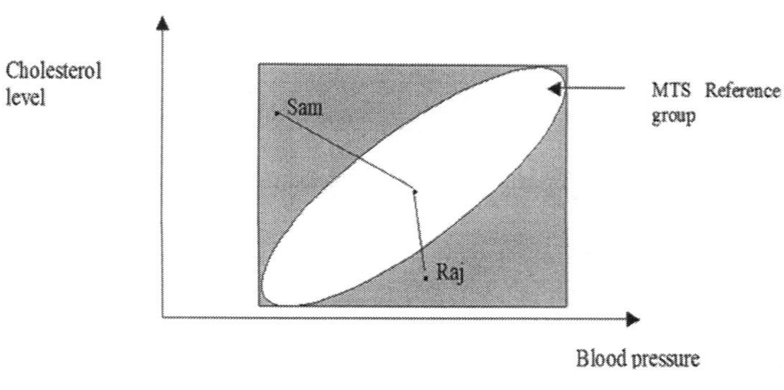

FIGURE 8.1 Importance of correlations in multivariate systems.

DOI: 10.1201/9781003190325-8

In Figure 8.1, if we look at the variables blood pressure and cholesterol levels independently of each other, then we can conclude that both Sam and Raj are normal (in the rectangular area). On the other hand, if we consider the correlations between the variables, we can say that they are somewhat abnormal, as they do not fall into the elliptical area, which forms the reference group. Also, we can precisely state that Sam is sicker than Raj, since his distance from the center of the ellipse is larger than that of Raj. Through MTS analysis, we can obtain such insights by expressing patterns in quantitative terms.

The Mahalanobis distance (MD), introduced by the well-known Indian statistician P.C. Mahalanobis, measures distances of points in multidimensional spaces. The Mahalanobis distance has been extensively used in several areas, such as spectrographic and agricultural applications. Because it takes correlations between the variables into account, the Mahalanobis distance is proven to be superior to other multidimensional distances, such as Euclidean distance. The Mahalanobis distance is calculated using the following equation:

$$\text{Mahalanobis distance} = (1/k)\, Z_i^T C^{-1} Z_i \tag{8.1}$$

where Z_i = standardized vector of X_i (i = 1..k); C = correlation matrix; k = no.of variables

The most important aspect of developing a multivariate measurement system using the Mahalanobis distance is having a reference point. While it is easier to obtain a reference point for a scale with a single characteristic, it is not so easy to obtain a single reference point when dealing with multiple characteristics. Therefore, in MTS, the reference point corresponding to multiple variables is obtained with the help of a group of observations that are as uniform as possible, yet still enable us to distinguish their different patterns through Mahalanobis distance. These observations are modified in such a way that their center is located at the origin, and the corresponding Mahalanobis distances are scaled so that the average distance of this group is unity. The set of observations used for the reference is often referred to as the "Mahalanobis space" (MS), "reference group," or "normal group." The selection of this group is entirely at the discretion of the decision-maker conducting the diagnosis. In fraud analytics, this group can be one where there are no incidences of fraud; in medical diagnosis, this group can be a set of people without any health problems; and in stock market predictions, this group could correspond to companies with steady average growth over a 3-year period. The observations in this group are similar, but not the same. Judicious selection of this group is extremely important for accurate diagnosis or predictions.

After developing the scale, the next step is the validation of the scale, which is done with the help of observations that are outside the Mahalanobis space. If the scale is good, the distances of these observations must align with the decision-maker's judgment. In other words, if an observation does not belong to the reference group, then it should have a larger distance. To quantify the diagnostic accuracy, we typically use the signal-to-noise (S/N) ratio metric. The S/N ratio captures the magnitude of real effects after making some adjustments for uncontrollable variation (noise) by taking into account the correlation between the true or observed information

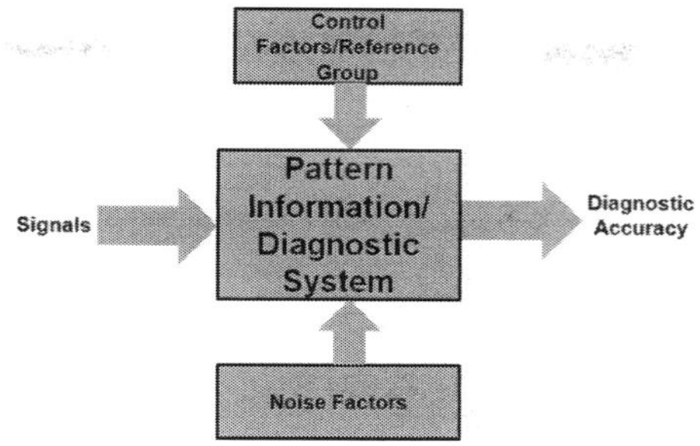

FIGURE 8.2 Pattern information or diagnostic system used in MTS.

(i.e., input signals) and the output of the diagnostic system (the Mahalanobis distance). In a multivariate context, the S/N ratio can simply be defined as a measure of the accuracy of diagnostics. A typical multidimensional predictive system used in MTS can be described using Figure 8.2:

As mentioned above, the output or diagnosis accuracy should have a good correlation with the input signal, and S/N ratios measure this correlation. The diagnostics are conducted based on the information on the variables defining the system and they should be "accurate" even in the presence of noise factors such as different places of measurement, operating conditions etc.

If the accuracy of the diagnosis is satisfactory, then we will identify a useful subset of variables that is sufficient for the measurement scale with acceptable diagnostic accuracy. Experience shows that in many cases, the accuracy with useful variables is better than that with the original set of variables. However, in some cases, the accuracy with useful variables might be lower—which might still be desirable as it helps reduce the cost of inspection or measurement. In multidimensional systems, the total number of combinations to be examined would be on the order of several thousand, and hence it is not possible to examine all combinations. Here, we propose the use of orthogonal arrays (OAs) to reduce the number of combinations to be tested. OAs are developed to estimate the effects of the variables by minimizing the number of combinations to be examined. They have been in use for quite a long time in the field of experimental design. In MTS, the variables are assigned to different columns of an OA. Based on the S/N ratios obtained from different variable combinations, important variables are identified. The future diagnosis is carried out only with these important variables.

8.3 STAGES IN MTS

From the above discussion, the basic stages in MTS can be summarized as follows:

Stage I: Develop Measurement Scale

- Select a Mahalanobis space or reference group with suitable variables and observations that are as much uniform as possible.
- Use the Mahalanobis space as a base or reference point of the scale.

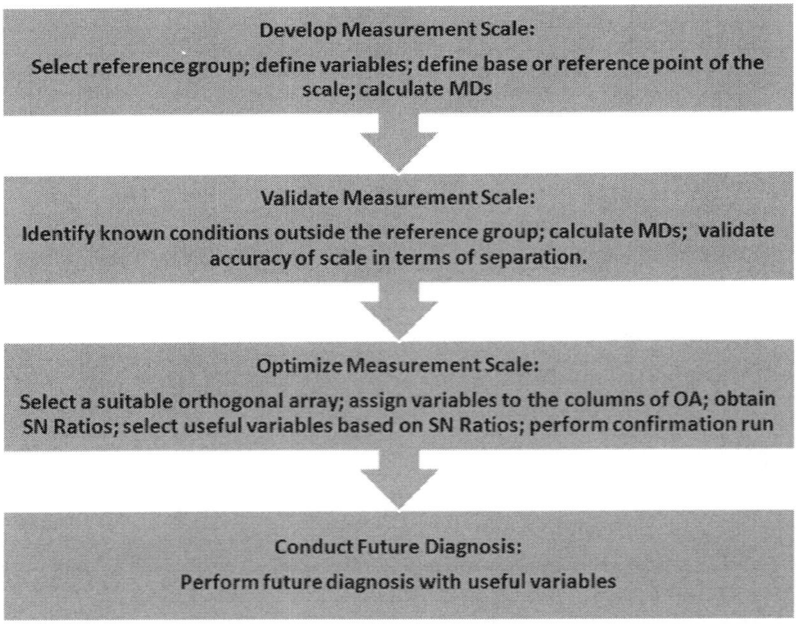

FIGURE 8.3 Steps in MTS.

Stage II: Validate Measurement Scale

- Identify the conditions outside the Mahalanobis space.
- Compute the Mahalanobis distances of these conditions and check if they match with decision-maker's judgment.

Stage III: Optimize Measurement Scale

- Optimize the scale by finding out the useful set of variables using orthogonal arrays and signal-to-noise ratios.

Stage IV: Conduct Future Diagnosis

- Monitor the conditions using the scale, which is developed with the help of the useful set of variables. Based on the values of Mahalanobis distances, take appropriate corrective actions.

Figure 8.3 shows different steps in MTS.

From the above discussion, it is clear that role of orthogonal arrays are prominent in the third stage of MTS analysis. Each experimental run in the orthogonal array design matrix uses a subset of variables; the resulting S/N ratios of these subsets are calculated using the distances outside of the reference group.

8.4 SIGNAL-TO-NOISE RATIO – A MEASURE OF DIAGNOSTICS

In the context of MTS, the Signal-to-Noise (S/N) ratio is used as a measure of diagnostic accuracy. The use of S/N ratios ensures a high level of prediction with useful variables. S/N ratios are computed for all combinations of the orthogonal array using Mahalanobis distances. S/N ratios are also used to select important variables for future diagnosis. Using S/N ratios as the response, average effects of variables are computed at level 1 (presence) and level 2 (absence). Based on these effects, the importance of variables can be determined. The examples provided in this chapter will show how such a determination can be made.

8.4.1 Types of S/N Ratios in MTS

In MTS application, typically the following types of S/N ratios are used:

- Larger-the-better type
- Nominal-the-best type
- Dynamic type.

When the true levels of abnormals are not known, larger-the-better type S/N ratios are used if all the observations outside reference group are abnormals. This is because the MDs for abnormals should be higher. If the observations outside reference group are a mixture of normals and abnormals then, nominal-the-best type S/N ratios are used. When the levels of abnormals are known, dynamic S/N ratios are used.

8.4.1.1 Equations for S/N Ratios

8.4.1.1.1 Larger the Better Type

The procedure for calculating S/N ratios corresponding to a run of an OA is as follows:

Let there be t abnormal conditions. Let D_3, D_2^2,.., D_t^2 be MDs corresponding to the abnormal situations. The S/N ratio (for larger the better criterion) corresponding to q^{th} run of OA is given by:

$$\text{S/N ratio} = \eta_q = -10\text{Log}_{10}\left[(1/t)(1/D_i^2)\right] \qquad (8.2)$$

8.4.1.1.2 Nominal-the-Best Type

The procedure for calculating S/N ratios corresponding to a run of an OA is as follows:

Let there be t abnormal conditions. Let D_3, D_2^2,.., D_t^2 be MDs corresponding to the abnormal situations. The S/N ratio (nominal-the-better type) corresponding to q^{th} run of OA is calculated as follows:

$$T = \text{Sum of all } D_i\text{'s} = \sum_{i=1}^{t} D_i$$

$$S_m = \text{Sum of squares due to mean} = T^2/t$$

$$V_e = \text{Mean square error} = \text{variance} = \sum_{i=1}^{t} \frac{(D_i - \bar{D})^2}{(t-1)}$$

Where $\bar{D}$, is average of D_i 's

$$\text{S/N ratio} = \eta_q = 10\text{Log}_{10}\left[\frac{1/n(S_m - V_e)}{V_e}\right] \tag{8.3}$$

8.4.1.1.3 Dynamic Type

Examples of this type include weather forecasting systems and the case of rainfall prediction. The procedure for calculating S/N ratios corresponding to a run of an OA is as follows:

Let there be t abnormal conditions. Let D_3, D_2^2,.., D_t^2 be MDs corresponding to the abnormal conditions. Let M_1, M_2,.., M_t be true levels of severity (rain fall values in rain-fall prediction system example).

$$S_T = \text{Total Sum of Squares} = \sum_{i=1}^{t} D_i^2,$$

$$r = \text{Sum of squares due to input signal} = \sum_{i=1}^{t} M_i^2,$$

$$S_\beta = \text{Sum of Squares due to Slope} = (1/r)\left[\sum_{i=1}^{t} M_i D_i\right]^2,$$

$$S_e = \text{Error Sum of Squares} = S_T - S_{\beta,}$$

$$V_e = \text{Error Variance} = S_e/(t-1)$$

The S/N ratio corresponding q^{th} run of OA is given by,

$$\text{S/N ratio} = \eta_q = 10 \log_{10}\{(1/r)\ [S_\beta - V_e]\ /V_e\} \tag{8.4}$$

8.5 MTS EXAMPLE

The different stages in MTS analysis are explained with the following example.

Let consider a case of four variables, X1, X2, X3 and X4. The raw data for "healthy group" with 15 observations and "abnormals" with 10 observations are given in Tables 8.1 and 8.2 respectively.

The above data is standardized with means and standard deviations of healthy group. Then correlation matrix is constructed by evaluating correlations between the variables. The correlation coefficient (ρ) between two variables X and Y is given by the following equation:

$$\rho = \frac{\text{Cov}(X,Y)}{V(X)V(Y)}$$

where, Cov (X,Y) = covariance between X and Y and

$$\text{Cov}(X,Y) = \frac{1}{n-1}\sum_{i=1}^{n}\left(Xi - \bar{X}\right)\left(Yi - \bar{Y}\right)$$

V(X) = variance of X and V(Y) = variance of Y

TABLE 8.1
Healthy Group Data

S.No	X1	X2	X3	X4
1	59	7	4.2	0.7
2	51	7	4.4	0.98
3	53	6.7	4.2	0.83
4	52	6.6	4.3	0.9
5	52	6.7	3.9	0.97
6	56	7.2	4	0.93
7	51	7.1	4	0.88
8	50	7.5	4.3	0.71
9	41	6.9	4.3	0.81
10	48	7	4.2	0.93
11	49	7.1	4	0.88
12	41	6.9	4.2	0.92
13	41	7.3	4.4	0.76
14	36	7.2	4	0.86
15	34	7.4	4.4	0.92

TABLE 8.2
Abnormal Data

S.No	X1	X2	X3	X4
1	30	8	4	0.59
2	46	7	3.3	0.48
3	22	7.2	3.8	0.52
4	31	7	3.8	0.57
5	33	6.7	3.4	0.51
6	38	7.1	3.8	0.61
7	34	8	3.7	0.62
8	42	6.7	3.2	0.54
9	30	7.4	3.9	0.34
10	28	7.3	3.6	0.53

The correlation coefficients are calculated and they are represented in correlation matrix C, which is given in Table 8.3.

The inverse matrix is obtained for this correlation matrix and is given in Table 8.4.

After obtaining inverse matrix, the Mahalanobis distances (MDs) are calculated based on equation 8.1. Tables 8.5 summarizes MDs for observations in healthy group and abnormals.

From Figure 8.4, it is clear that by using the MTS scale, there is a good separation between normals and abnormals. Based on this validation process, we can safely say

TABLE 8.3
Correlation Matrix

	X_1	X_2	X_3	X_4
X_1	1.000	-0.368	-0.289	-0.088
X_2	-0.368	1.000	0.194	-0.339
X_3	-0.289	0.194	1.000	-0.262
X_4	-0.088	-0.339	-0.262	1.000

TABLE 8.4
Inverse Matrix

	X_1	X_2	X_3	X_4
X_1	1.357	0.564	0.390	0.413
X_2	0.564	1.380	0.033	0.526
X_3	0.390	0.033	1.200	0.360
X_4	0.413	0.526	0.360	1.309

that the scale is good. The next stage is the optimization or screening stage. In this stage, a suitable orthogonal array is used to test different combinations of the variables. Since there are four variables, the L8(27) array is selected for optimization. Some other examples of orthogonal arrays are provided in Appendix II. The four variables are assigned to the first four columns of the array. For all combinations, abnormal MDs are obtained, and S/N ratios are calculated based on these MDs. Table 8.6 summarizes the results of various combinations.

From the S/N ratios, average responses are computed at level 1 (presence) and level 2 (absence) of the variables. Figure 8.5 shows average response curves. From this figure, it is clear that except X2 all other variables are important for separating normals and abnormals. With X1, X3 and X4 a conformation run was conducted and it is found that separation is good with these useful variables. In fact, separation after optimization is better than initial separation. The results of confirmation are given in Table 8.7 and Figure 8.6. Therefore, we can conclude that these variables are sufficient for future diagnosis.

8.6 IMPROVING CLIENT EXPERIENCE CASE STUDY

This project focused on improving the measurement of key indicators of client relationship health at a financial company. The health of a client was determined based on 49 variables, which were categorized into green (loyal), yellow (vulnerable), and red (at risk) groups of clients. The existing practice was to make decisions based on the client relationship manager's assessment. There was no quantitative approach in place. Since the method was subjective, it was not able to identify or differentiate specific types of risk from the primary client risk factors. Therefore, the project's primary goal was to develop a metric to determine the client's health quantitatively.

TABLE 8.5
MDs for Healthy Group and Abnormals

S.No	Healthy	Abnormal
1	1.4016	7.3791
2	1.5880	19.4915
3	0.6447	15.7426
4	0.9995	11.3034
5	1.0397	23.0483
6	1.2154	7.2853
7	0.3880	9.7237
8	1.3582	22.7003
9	0.7886	17.7513
10	0.1772	15.7460
11	0.3589	
12	0.4335	
13	0.7547	
14	1.4309	
15	1.4210	

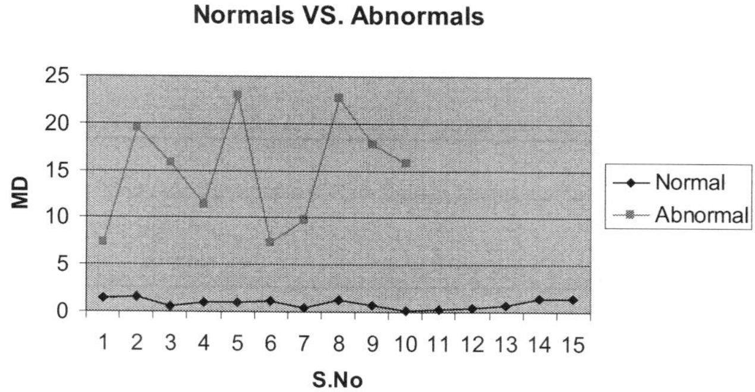

FIGURE 8.4 Separation Between Normals and Abnormals.

The MTS methodology was applied to quantitatively determine client health. With all 49 variables, the MTS scale was constructed and validated. These are shown in Figure 8.7.

After performing MTS optimization with L_{64} OA, the results are as shown in Table 8.8

After performing SN ratio analysis and considering other practical constraints, 11 variables were selected to be important. With these 11 variables, the performance of the scale was better than the scale with 49 variables in terms of separation. This is clear from Figure 8.8.

TABLE 8.6
Optimization with $L_8(2^7)$ Array

	X1	X2	X3	X4				Abnormal MDS										
Run	1	2	3	4	5	6	7	1	2	3	4	5	6	7	8	9	10	S/N Ratio
1	1	1	1	1	1	1	1	7.38	19.49	15.74	11.30	23.05	7.29	9.72	22.70	17.75	15.75	11.04
2	1	1	1	2	2	2	2	6.61	10.36	8.10	5.08	12.27	3.17	9.80	14.03	4.23	9.05	8.28
3	1	2	2	1	1	2	2	8.59	9.90	15.26	9.03	11.05	5.51	6.17	7.53	22.40	11.94	9.63
4	1	2	2	2	2	1	1	5.79	0.05	12.24	5.15	3.98	1.72	3.45	0.59	5.79	7.17	-3.89
5	2	1	2	1	2	1	2	8.91	11.25	8.10	6.66	12.36	4.59	8.25	10.67	18.29	7.41	9.29
6	2	1	2	2	1	2	1	13.78	0.02	0.38	0.02	1.73	0.05	13.78	1.73	1.94	1.01	-10.32
7	2	2	1	1	2	2	1	6.96	31.81	13.75	11.13	25.90	9.29	10.98	31.97	23.81	18.16	11.53
8	2	2	1	2	1	1	2	1.23	27.70	5.27	5.27	21.80	5.27	8.35	34.30	2.90	12.13	6.90

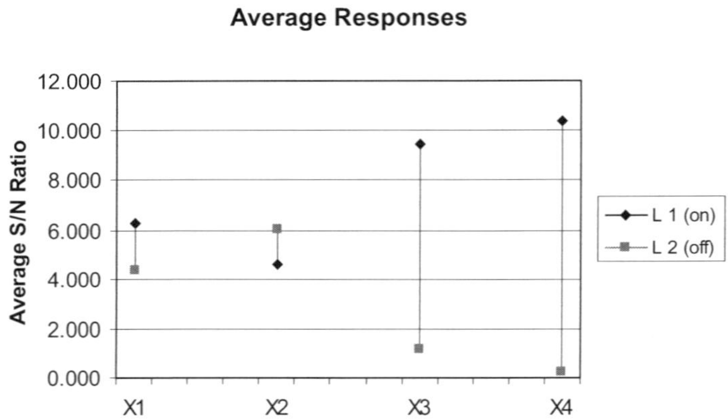

FIGURE 8.5 Average Response Curves.

8.6.1 Improvements Made Based on Recommendations from MTS Analysis

A new measurement system has been developed by including the 11 attributes from the following:

- Survey
- Volatility within client
- Volatility within company

The company convinced that the performance of the measurement system with these 11 attributes is much is more meaningful and objective as compared to traditional subjective evaluations.

TABLE 8.7
MDs from Confirmation Run

S.No	Normal	Abnormal
1	1.8431	8.8017
2	1.9700	24.0530
3	0.2352	18.4000
4	0.5695	12.3652
5	1.1897	24.1142
6	0.7906	8.7156
7	0.4203	11.4534
8	1.0899	25.5928
9	0.4587	21.9678
10	0.2257	19.7634
11	0.4230	
12	0.3716	
13	0.9863	
14	1.9050	
15	1.5213	

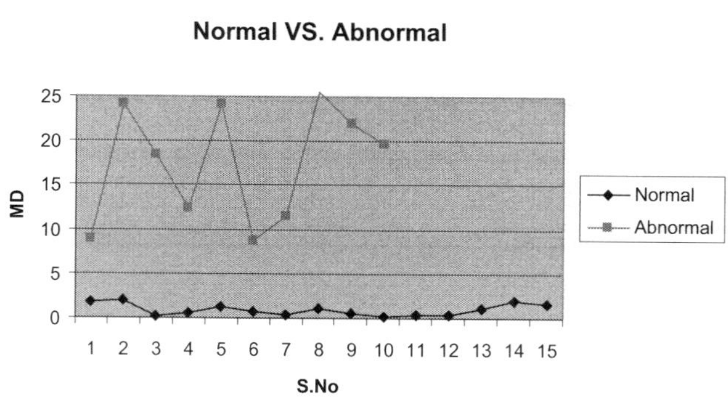

FIGURE 8.6 Separation Between Normals and Abnormals.

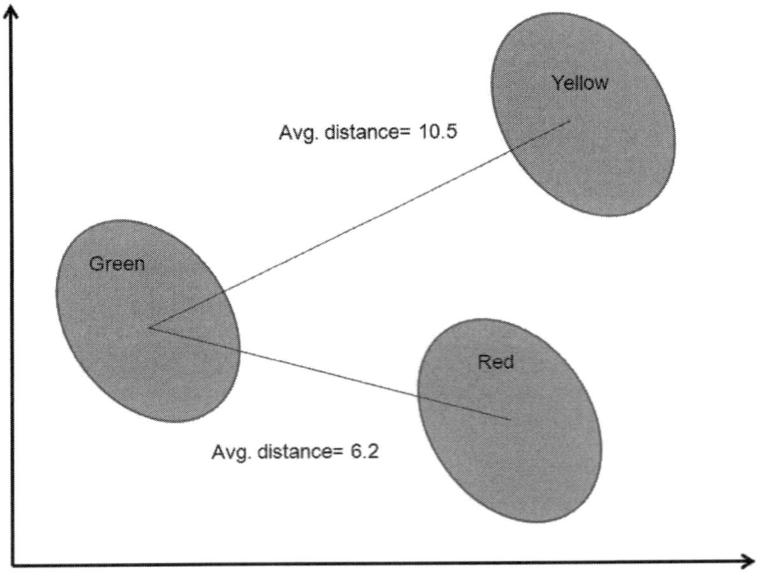

FIGURE 8.7 Validation of Scale – distance between green, yellow and red clients (with 49 variables).

TABLE 8.8
MTS Optimization with L_{64} Array

Run #	S/N Ratio	Run #	S/N Ratio	Run #	S/N Ratio	Run #	S/N Ratio	Run #	S/N Ratio	Run #	S/N Ratio	Run #	S/N Ratio
1	0.65	11	0.26	21	0.68	31	0.69	41	0.04	51	0.03	61	0.92
2	1.22	12	-0.34	22	0.77	32	0.80	42	0.20	52	0.11	62	0.82
3	0.64	13	0.67	23	0.85	33	0.17	43	-0.13	53	0.60	63	1.04
4	0.04	14	0.70	24	0.87	34	-0.20	44	-0.43	54	0.68	64	0.93
5	1.09	15	0.81	25	0.12	35	0.03	45	0.51	55	0.65		
6	1.12	16	0.89	26	0.04	36	0.45	46	0.83	56	0.75		
7	0.47	17	-0.30	27	0.03	37	0.80	47	0.56	57	-0.06		
8	0.60	18	-0.30	28	0.22	38	0.79	48	0.82	58	0.42		
9	-0.23	19	-0.10	29	0.82	39	0.93	49	0.42	59	0.28		
10	0.32	20	0.36	30	0.85	40	0.91	50	-0.36	60	0.04		

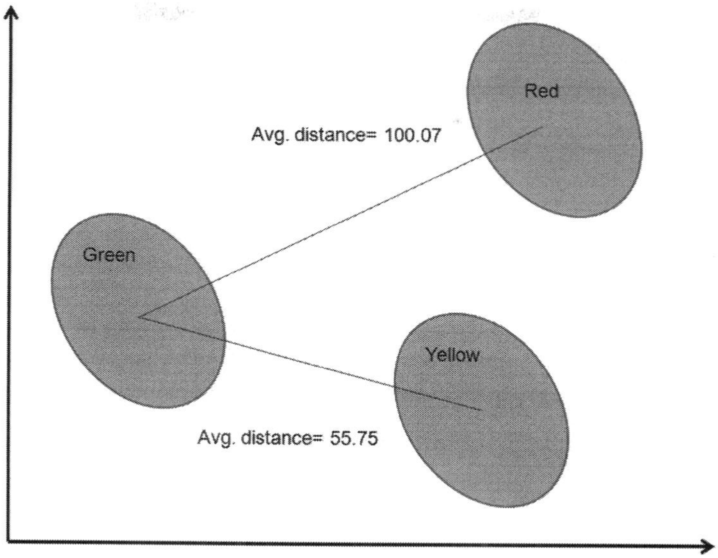

FIGURE 8.8 Separation between clients (after optimization, with 11 variables).

NOTE

1 Thanks are due to Wiley and ASQ Quality Press for allowing us to use materials from prior publications of one of the authors.

REFERENCES

Genichi Taguchi and Rajesh Jugulum, 2002. The Mahalanobis-Taguchi Strategy: A Pattern Technology, John Wiley & sons.

Phadke, Madhav S. 1989. Quality Engineering Using Robust Design, Prentice Hall, New Jersey, Englewood Cliffs.

Taguchi, G., 1987. System of Experimental Design, Vols. 1 and 2, ASI & Quality Resources, White Plains, New York.

Taguchi, G., R. Jugulum, R. and S. Taguchi, 2004. Computer Based Robust Engineering: Essentials for DFSS, ASQ quality press, Milwaukee, WI.

9 Big Data and Artificial Intelligence (AI)

9.1 INTRODUCTION

With the advent of computational breakthroughs and human and machine-generated data, Big Data and Artificial Intelligence (AI) have become complementary in their applicability and value creation. In addition to machine generated data and the data from the transactional systems, user actions on the internet and various events leave a digital fingerprint, creating an ever-increasing reservoir of data called Big Data. Such a rich repository is valuable for performing advanced analytics and AI-driven insights for decision-making and innovation.

AI is the ability of a machine to perform cognitive functions we associate with human minds, such as perceiving, reasoning, learning, and problem-solving. Machine Learning (ML) detects patterns and learns how to make predictions and recommendations by processing experiences, rather than by explicit programming. Deep Learning (DL) is a subset of ML in which the tasks are broken down and distributed onto ML algorithms that are organized in consecutive layers. Each layer builds upon the output from the previous layer. Together, the layers constitute an artificial neural network (ANNs) that mimics the distributed approach to problem-solving carried out by neurons skin to the human brain.

Big Data is a treasure trove for uncovering hidden patterns and trends. However, to decipher these insights, one needs scalable AI tools and methods. Big Data provides data for creating actionable intelligence, drive progress across various field - business, science and society. Integrating data from multiple sources uncovers tremendous value and offers insights that were not possible earlier. The use cases of Big Data are extensively documented in the last decade.

AI acts as an intelligent navigator. It uses advanced algorithms and computational power to sieve through massive datasets, uncover patterns, predict trends, and extract actionable knowledge. Big Data and AI form a powerful synergy that is revolutionizing the way people live, work, and interact with the world.

9.2 THE UNREASONABLE EFFECTIVENESS OF DATA

"The Unreasonable Effectiveness of Data" is an influential paper published by Google researchers and has been a topic of interest among researchers in the field of Machine Learning (ML) and Artificial Intelligence (AI). A central thesis of the paper is that, the quantity of data often trumps the choice of the model in ML tasks. This is supported by findings that as the size of the dataset increases, the performance differences between various ML models diminish. The paper suggests that even simple models can achieve high performance when trained on large datasets

DOI: 10.1201/9781003190325-9

and that, with enough data, complex problem-solving can be reduced to essentially a form of memorization.

The paper has garnered significant attention for its insights into the role of data in ML, particularly in fields related to Natural Language Processing (NLP) and Computer Vision (CV). Google's research team has conducted experiments with an internal dataset of 300 million images labeled with over 18,000 categories, known as JFT-300M. Their findings indicate a logarithmic relationship between the amount of training data and performance on vision tasks, suggesting that performance improvements may continue as datasets grow even larger. These results have led to new state-of-the-art outcomes in various benchmarks, reinforcing the importance of large-scale datasets for visual pre-training.

The concept echoes Eugene Wigner's earlier observations in "The Unreasonable Effectiveness of Mathematics in the Natural Sciences," which questioned why mathematics is so adept at explaining physical phenomena. The effectiveness of data in ML challenges philosophies that do not accept that science is fundamentally about discerning patterns of experience.

The research community is encouraged to focus on obtaining large-scale, task-specific data to advance the field further. There is an ongoing debate about whether creating even larger datasets, such as those with over a billion images, will continue to yield meaningful improvements. The paper's philosophy aligns with the view that embracing complexity and leveraging large datasets is more beneficial than striving for overly elegant theories, especially in sciences involving human behavior.

"The Unreasonable Effectiveness of Data" underscores the pivotal role of large datasets in advancing ML and AI. It suggests that the sheer volume of data can compensate for simpler models and that the pursuit of ever-larger datasets may continue to push the boundaries of what is achievable in the field. As the community moves forward, the focus on building and utilizing massive datasets will likely remain a central theme in the quest for improved ML/AI systems.

9.3 DATA NECESSITY AND SUFFICIENCY

In the context of AI/ML models, data necessity and sufficiency are critical aspects of data that determine the effectiveness of these systems. Data necessity refers to ensuring that the collected data is relevant and aligned with the model's objectives, meaning only the most pertinent features or variables are included. This reduces noise and computational complexity. On the other hand, data sufficiency ensures that the dataset is comprehensive and diverse enough to represent the full range of scenarios the model will encounter in real-world applications. A model trained on insufficient data risks becoming biased or overfitting, limiting its generalizability. Addressing both aspects allows for the development of robust, high-performing AI models capable of accurate predictions and insights (Figure 9.1).

9.3.1 DATA NECESSITY

Data necessity is an aspect of data that can provide necessary features for modeling tasks in ML It ensures that the dataset contains all relevant features needed to

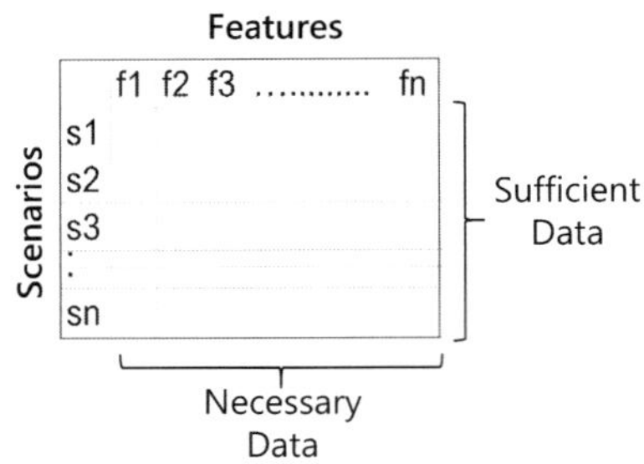

FIGURE 9.1 Data in a table where rows represent sufficiency and columns necessity.

perform tasks such as prediction, classification, or clustering accurately. Relevant features significantly impact the outcome variable and contribute to the model's ability to learn patterns effectively.

Sometimes, necessary features are not directly available in the raw data. In such cases, features are engineered, which involves creating new features from existing ones to capture the underlying patterns in the data.

While having the necessary features is essential, having redundant features can lead to overfitting and increased computational complexity. Techniques like Principal Component Analysis (PCA) or feature selection methods help identify and retain only the most necessary features, balancing complexity with performance.

Understanding the domain from which the data is derived is essential. It helps identify necessary features. Domain experts can provide insights into the required features, guiding the selection process.

Having necessary features also aids in model interpretability. When models are built using relevant and essential features, it becomes easier to understand how predictions are made and why certain decisions are reached. This is crucial for trust and transparency in AI systems.

9.3.2 Data Sufficiency

Data sufficiency, on the other hand, focuses on having enough data to allow the model to generalize well. Sufficient data helps the model identify patterns and rules that apply broadly rather than overfitting the specific examples in the training set. The model should have enough data to learn generalizable patterns rather than memorizing specific examples.

The goal is typically to have representative, diverse, and abundant data to train models that generalize well to new, unseen data within the problem domain.

In AI and ML, ensuring data sufficiency involves curating datasets that comprehensively capture the variations in the real-world environment where the model will operate. This means collecting diverse examples and accounting for edge cases, rare occurrences, and potential anomalies. A dataset with sufficient coverage helps prevent overfitting to a narrow distribution and allows the model to generalize better to unseen data.

However, practically achieving complete sufficiency is difficult due to the infinite variability in real-world data. We should typically aim for:

- **Representative Coverage:** The data should cover a wide range of important scenarios or cases that the model will likely encounter in its intended application.
- **Edge Case Inclusion**: While not covering every possible scenario, the data should include important edge cases or rare events critical to the model's performance.
- **Balanced Distribution**: The data should represent different classes or categories in a balanced way, unless the real-world distribution is intentionally imbalanced.
- **Comprehensive Coverage**: A dataset encompassing many scenarios ensures the model can learn from diverse examples. This helps the model effectively handle edge cases and unexpected inputs.
- **Bias Reduction**: Ensuring data necessity helps minimize bias. If certain scenarios are underrepresented, the model might perform poorly on them, leading to biased outcomes.
- **Robustness:** Models trained on necessary data are generally more robust and reliable because they have been exposed to various situations during training.

Sufficiency can be evaluated using tools like learning curves, which plot model performance as a function of dataset size, helping identify when adding more data no longer results in significant performance gains. Models trained on sufficient data tend to perform better, as they have a solid foundation to learn from and can make accurate predictions on unseen data. Data sufficiency is often determined empirically by observing when additional data no longer significantly improves model performance.

- **Statistical Power:** There should be enough data to draw statistically significant conclusions and build reliable models. With sufficient data, the model's predictions are more statistically significant, reducing the likelihood of random noise affecting the learning process.
- **Complexity Matching**: The amount of data should be sufficient given the trained model's complexity. More complex models typically require more data to avoid overfitting.

When discussing data in ML and AI, we often also consider Data Quality (accuracy, consistency, and reliability of the data), Data Diversity (variety of data sources and types represented), and Data Relevance (how well the data aligns with addressing a specific problem). Data sufficiency addresses the quantity and quality of data needed for a model to capture the underlying patterns and relationships in a dataset.

Balancing data necessity and sufficiency is a key challenge in ML. Ensuring that data is both necessary and sufficient is crucial for developing accurate, unbiased, and generalizable models. Balancing these aspects requires careful consideration of data collection, preprocessing, and model selection strategies.

- **Data Collection**: Gathering necessary and sufficient data can be resource-intensive, but it is essential for building effective models.
- **Data Augmentation**: Techniques such as data augmentation can help simulate additional scenarios, contributing to necessity and sufficiency.
- **Transfer Learning**: Leveraging pre-trained models can help address data sufficiency by allowing models to benefit from previously learned knowledge.

9.4 DATA CLEANLINESS

Data cleanliness is foundational to the success of AI applications, influencing their accuracy, fairness, efficiency, and user trust. Ensuring high data quality through proper data cleansing processes is essential for optimizing AI performance and achieving reliable outcomes. Data cleanliness significantly impacts AI performance in several ways:

- **Accuracy and Reliability**: Clean data ensures that AI models are trained on accurate and consistent information, leading to more reliable predictions. Errors, inconsistencies, or biases in the data can lead to flawed outcomes, such as false positives or negatives, undermining the model's accuracy and reliability.
- **Bias Mitigation**: Clean data helps identify and rectify biases, ensuring that AI systems are fair and unbiased. If the training data is biased, the AI/ML models may amplify or replicate these biases, resulting in unfair or discriminatory outcomes.
- **Generalization and Robustness**: AI models need to generalize well to unseen data. Clean training data representing a wide range of scenarios helps AI systems handle diverse inputs and make accurate predictions, thus enhancing their robustness.
- **Efficiency and Performance**: Models trained on clean data require less computational power and time to process information, improving performance and scalability. This efficiency is critical, as it reduces resource requirements and enhances the overall performance of AI solutions.
- **User Trust and Adoption**: Consistently accurate outcomes foster trust in AI systems. Clean data ensures that AI delivers reliable results, encouraging user adoption and building stronger relationships with customers and stakeholders.

9.5 HANDLING INCOMPLETE DATA

Organizations can identify and address incomplete data in their datasets by employing several strategies and techniques:

- **Understanding the Nature of Missing Data**: Missing data can be categorized into three types:
  - Missing Completely at Random (MCAR)

- Missing at Random (MAR)
- Missing Not at Random (MNAR)

  Each type requires different handling strategies. For instance, MCAR data can be safely removed without introducing bias, while MAR and MNAR data may require more sophisticated handling techniques.
- **Data Audits and Monitoring**: Regular data audits can help identify patterns of missing data. By continuously monitoring data collection processes, organizations can detect when and where data gaps occur and take proactive measures to address them.
- **Imputation Techniques**: Imputation involves replacing missing values with estimated ones. Standard methods include using the mean, median, or mode for numerical data and the most frequent category for categorical data. More advanced techniques, like k-nearest neighbors (KNN) or regression imputation, can estimate missing values based on relationships within the data.
- **Data Removal**: In some cases, especially when data is MCAR, removing incomplete data points may be appropriate. However, this should be done cautiously to avoid reducing the dataset size too much, which could lead to biased results.
- **Data Collection Improvements**: Improving data collection processes can prevent missing data. This includes automating data entry to reduce human error, ensuring sensors and systems function correctly, and establishing clear guidelines for data collection.
- **Data Governance and Quality Management**: Implementing robust data governance frameworks can help maintain data quality. This includes defining policies and standards for data management, ensuring consistent data collection and storage practices, and regularly reviewing data quality.

9.6 BIG DATA AND AI

Big Data refers to vast collections of data gathered from diverse sources such as sensor readings, social media interactions, and business operations, often in various formats. In contrast, AI encompasses techniques and methods that enable machines to augment human intelligence. AI algorithms are trained on large datasets to identify patterns and make predictions. However, AI comes with certain limitations:

- **Data Dependence**: AI's effectiveness relies heavily on the quality and volume of data. Inadequate or inaccurate data can lead to faulty or unreliable model outputs, often called the "garbage in, garbage out" problem.
- **Explainability**: The inner workings of complex AI models are generally opaque and difficult to interpret or understand, making it challenging to explain how they arrive at their conclusions. Such models are often called "black box" systems, where the reasoning behind their outputs is not easily interpretable.

- **Bias**: Biases in the datasets used to train AI models can be replicated, leading to unfair or discriminatory outcomes in the AI system's decisions and predictions.

The effectiveness of AI depends on Big Data; together, they create significant value. Big Data supplies the fuel - vast amounts of data, while AI serves as the engine, processing this data through advanced algorithms to enable organizations to:

- Analyze large datasets to uncover trends, correlations, relationships, and patterns that would be impossible to detect using traditional/manual methods.
- Make data-driven decisions based on empirical evidence (learning from historical data) and real-world experiences, which will lead to greater efficiency and accuracy in decision-making.
- Automate complex tasks and processes, freeing up human resources to focus on more strategic, high-level initiatives.

Imagine a detective with countless clues but no way to analyze them and an expert analyst lacking the data needed to work. This analogy captures the individual limitations of Big Data and AI. -both are interdependent for creating value.

Despite being a trove of potential insights, Big Data presents challenges due to its sheer volume and complexity. Meanwhile, AI, though equipped with advanced algorithms, requires high-quality data to learn and operate effectively. Big Data fuels the rich datasets that AI algorithms traverse to uncover deep patterns and insights. Together, they create a powerful feedback loop— AI continuously learns from Big Data, refining its ability to analyze and predict while generating new data to enhance future data collection and model performance. Their complementarity lies in Big Data's ability to supply the massive quantities of structured and unstructured information that AI systems depend on for training and operation (Figure 9.2).

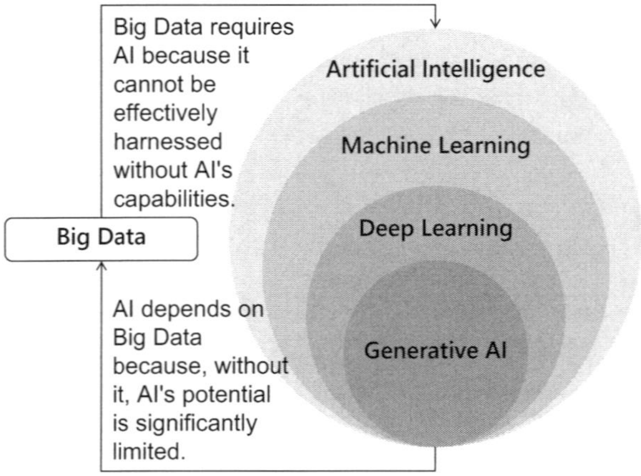

FIGURE 9.2 Symbiotic relationship between Big Data and AI.

Big Data is the foundation upon which AI models build their understanding of complex scenarios, enabling them to deliver decisions, predictions, and recommendations. This is the essence of the symbiotic relationship between Big Data and AI.

9.7 DATA ENGINEERING AND BIG DATA

Data engineering is the backbone of Big Data management. Data engineers develop, maintain, and optimize Big Data infrastructure, which includes designing data architecture, constructing efficient data pipelines, and ensuring the quality and accessibility of data for analytics.

Data engineering involves adopting scalable data architecture and storage mechanisms, and designing and operationalizing data pipelines that automatically move data from various sources into data warehouses or any of the many modern data architectures, such as data lakes.

The ETL processes ensure that the data is clean, accurate, and ready to be analyzed. They involve integrating data from various sources into usable formats for data scientists and analysts and managing batch and real-time data processing workflows.

Data engineering is essential for transforming raw, unstructured Big Data into a structured, clean, and accessible format that AI/ML models can leverage to generate actionable insights.

AI/ML models rely heavily on the quality, structure, and accessibility of data to perform tasks like pattern recognition, trend prediction, and decision-making. Data engineering ensures that this data is properly collected, processed, and stored through robust data pipelines, optimized architectures, and ETL processes. Without adequate data engineering, AI/ML algorithms may struggle with poor data quality, missing values, or inefficiencies in data retrieval, all of which can hinder their ability to deliver meaningful insights.

In short, data engineering lays the foundation for AI/ML to function optimally, turning vast amounts of Big Data into valuable assets.

9.8 BIG DATA AND ALGORITHMIC COMPLEXITY

The "Unreasonable Effectiveness of Data" remains relevant even in the context of AI/ML models developed using curated datasets that meet data sufficiency and necessity requirements. Large amounts of data can often compensate for less sophisticated algorithms, especially in domains like NLP and CV, where the richness and variability of data can capture complex patterns and nuances. Despite advancements in data curation and model development, the principle still holds because:

- **Data Abundance vs. Model Complexity**: The paper argues that having access to vast amounts of data often leads to better performance than relying solely on complex models with limited data. This is particularly true when the data captures rare human behavior or language aspects that sophisticated models might miss without extensive training data.
- **Generalization from Large Datasets**: Generalizing from a large dataset remains critical, even with curated datasets. Large datasets are effective

because they can provide examples across a wide range of scenarios, which helps models generalize better to unseen data.
- **Challenges with Data Curation**: While curated datasets are valuable, they are often limited by the resources required for their creation and maintenance. Even if noisy, large-scale datasets can provide useful insights that curated datasets might miss due to their limited scope.

While properly curated datasets are essential for developing effective AI/ML models, the "Unreasonable Effectiveness of Data" remains relevant because large datasets provide a broader base for learning and generalization.

9.9 GENERATIVE AI (GEN AI)

AI is a technology used to approximate—often to transcend—human intelligence and ingenuity through software and systems. Computers using AI are programmed to carry out highly complex tasks and analyze vast amounts of data. AI systems process historical data to detect patterns, improve decision-making, eliminate manually intensive tasks, and improve business outcomes. AI can

- **Isolate Patterns**: AI can spot patterns among vast amounts of data. It does this using specialized GPU processors that enable super-fast computing. Some systems can predict how those patterns might impact the future or help perform "what-if" analyses; this is called predictive analytics.
- **Improve Data Analytics**: AI adds another dimension to data analytics. AI increases the chance of success and achieving positive outcomes by basing data analytics decisions on a much wider volume of data (ideally higher-quality data), whether historical or real-time. By rapidly detecting data patterns, business processes can be improved to bring about better business outcomes.
- **Assist Better Decisions**: AI can provide opportunities such as product recommendations, responsive customer service, and tighter management of inventory levels. Some executives use AI as an "additional advisor," meaning they incorporate recommendations from their colleagues and AI systems, weighing them accordingly.

ML Algorithms can be regarded as the essential building blocks of AI. AI uses various coordinated algorithms to discover a signal among the noise of a mountain of data and find paths to solutions that emulate human cognition and understanding.

AI is a general-purpose technology. It has almost limitless use cases – and more seem to crop up weekly – with use cases across various industries.

9.9.1 Gen AI Capabilities

AI processes data and returns expected results, such as analyses or predictions; Gen AI produces newly synthesized content, like text, images, and sound, based on training from corpora of existing data. The reason Gen AI became so popular so quickly is that it empowers the end user. Gen AI enables individuals with varying technical expertise to interact with complex datasets, thus democratizing data accessibility. It empowers non-technical stakeholders to derive meaningful insights without specialized skills. Right now, anyone can log on to services like ChatGPT, Claude, Microsoft Copilot, or Gemini and start using them without any training. Gen AI has brought AI to the user level! There are no barriers to entry for what/which/who users can use it.

AI necessitates rigorous data preparation and processes to develop and test a model designed to produce a good outcome. With generative AI, you can start talking to it, and it will understand what you want to do and offer a response.

AI is a technology that can perform tasks that typically require human intelligence. It is often used to build systems with the cognitive capacity to mine data and continuously boost performance—to learn—from repeated events. AI is focused on detecting patterns, making predictions, automation, and generating insights. AI deployments help improve decision-making, remove the tedium from repetitive tasks, or spot anomalies (Figure 9.3).

Gen AI, in contrast, is a specific form of AI designed to generate content. This content could be text, images, video and music. Gen AI starts with a prompt that lets a user submit a question and any relevant data to guide content generation. Generative AI uses complex algorithms, deep learning, and large language models to generate new content based on the data it is trained on. It is a specific and narrow application of AI to creative use cases. It uses AI algorithms to analyze patterns in datasets to mimic style or structure and replicate different types of content.

Gen AI combines algorithms, large language models, and deep learning/neural network techniques to generate content based on patterns observed in other content. Gen AI outputs may appear new and original, but they are fundamentally the product of ML and other AI techniques that draw upon others' prior creativity. These systems leverage existing content repositories, often encompassing significant portions of the internet, to mimic human creativity. By repeatedly processing and reshaping this existing material, generative AI transforms it into a flexible data source capable of

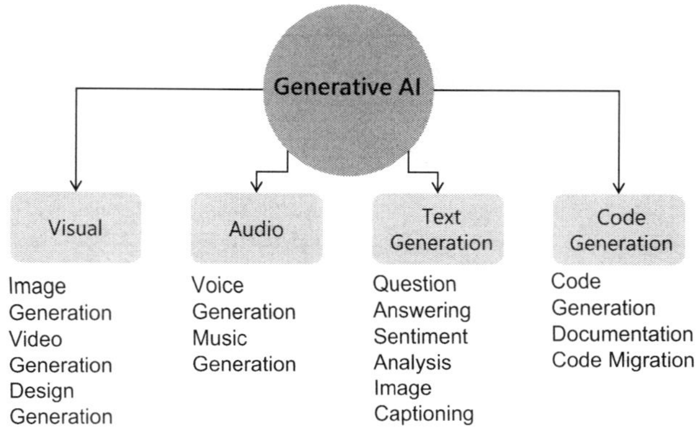

FIGURE 9.3 Capabilities of Generative AI.

producing content in response to user prompts. It synthesizes "new" outputs from the foundation of previously created works.

- **Generate Text**: They can generate legible text on various topics. They can compose business communication, draft articles/blogs, and even compose a complete annual report. Some journalistic organizations have experimented with generative AI programs to create news articles.
- **Generate Images**: They can generate realistic or surreal images from text prompts, create new scenes, and simulate a new painting. Note, however, that the fact that these images are based initially on the images fed into the Gen AI system is prompting lawsuits by creative artists (not only graphic artists but writers and musicians as well).
- **Generate Video**: It can automatically compile video content from text and put together short videos using existing images.
- **Generate Music**: It can compile new musical content by analyzing a music catalog and rendering a similar composition in that style. While this has caused copyright issues (as noted in the Drake and The Weeknd example above), generative AI can also collaborate with human musicians to produce fresh and arguably interesting new music.

To reemphasize, Gen AI is a specific use case for AI that is used for sophisticated modeling with a creative goal. It takes existing patterns and combines them to generate something new. Because of its creativity, Gen AI is seen as the most disruptive form of AI.

9.9.2 Large Language Models (LLMs)

LLMs are a class of Gen AI models designed to understand and generate human-like text based on the patterns and structures they've learned from massive datasets. LLMs predict the next word or phrase in a sequence by analyzing relationships between words in large datasets, a process known as unsupervised learning. These models are typically built using deep learning techniques, particularly transformer architectures, and are trained on extensive amounts of text data sourced from books, websites, articles, and other written content.

LLMs have scale (billions of parameters), making them capable of capturing complex language patterns. They can understand context, generate coherent responses, continue conversations, answer questions, or write essays. LLMs are not task-specific; they can perform various functions, such as language translation, summarization, content creation, and even coding. They generate text based on statistical patterns learned from data without understanding or reasoning.

LLMs have become a disruption by enabling end-users to interact with AI. These models' effectiveness heavily depends on the quality of the data on which they are trained. As discussed earlier, data quality is critical in developing AI systems, serving as the foundation for models to learn and make decisions. Data quality dimensions

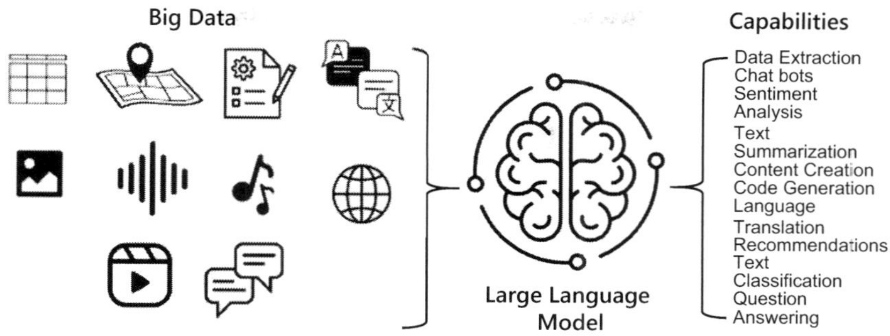

FIGURE 9.4 LLMs are driven by Big Data.

such as accuracy, completeness, relevance, and consistency are crucial for robust LLMs. Poor quality or biased data can result in unpredictable or ineffective models, eroding user trust and limiting applications. As LLMs evolve and integrate into various applications and workflows, the emphasis on data quality becomes increasingly important (Figure 9.4).

The training process for LLMs often involves using a vast, uncurated mix of high-quality, mediocre, and unreliable datasets from the internet, which can lead to concerns about the credibility and reliability of their results. The vast, unfiltered real-world texts used to train LLMs contain inherent biases and inaccuracies, which can lead to the generation of false information. The concept of 'textbook quality' data, which means clear, independent, instructive, and balanced data, is beneficial for training LLMs.

The credibility of results produced by LLMs is indeed a concern. Issues such as hallucination, biases, inaccuracies, and security vulnerabilities can compromise the reliability of these models. Rigorous data curation, ethical practices, and continuous security measures are essential to mitigate these risks. As the technology evolves, so must the strategies to ensure that LLMs produce credible and trustworthy outputs. Attribution methodologies are being refined to enhance the reliability and veracity of responses generated by LLMs. Leveraging external knowledge sources for attribution can help reduce hallucination and improve the truthfulness of generated text.

9.9.3 LLM Limitations

LLMs are used in various applications, including copywriting, knowledge base answering, text classification, code generation, and text generation. They demonstrate flexibility and potential in multiple tasks. They come with the following limitations as well:

- **Hallucinations and Misleading Outputs**: LLMs can produce outputs that sound plausible, confident, and authoritative—outputs that are entirely fabricated, potentially leading to misleading or factually incorrect results. This phenomenon is known as "hallucination". Users can also manipulate LLMs (through prompting) to return harmful information by altering text inputs, which raises concerns about the models' susceptibility to producing unreliable result.

- **Biases in Outputs**: LLMs may inadvertently generate biased content due to the data they are trained on, reflecting societal biases. This can result in the generation of racist, sexist, or otherwise harmful comments.
- **Toxic or Harmful Speech**: Some LLMs can occasionally output toxic or harmful speech, although major providers are working on improving the safety of these models to mitigate such outputs.
- **Recency of Knowledge**: LLMs can make mistakes or generate false information because they cannot always access the most recent facts. This limitation necessitates continuous adjustments and the use of additional tools to ensure the accuracy of the information provided.
- **Limitations of Understanding/Reasoning**: LLMs are trained on vast text corpora and learn the relationships between words in a given context. However, they lack a system to connect words with the concepts associated with tangible objects and real-world occurrences. This gap prevents LLMs from truly understanding the text's meaning and intent/context. They may also face difficulties with tasks that require mathematical calculations or other forms of computation.
- **Notion of Time, Space, and Causality**: Humans have a strong understanding of time, space, and causality, which shapes our perception of the world. LLMs, however, lack mechanisms to incorporate these constructs, resulting in outputs strictly driven by the text seen during training.

9.9.4 Data Quality for Building LLMs

LLMs have shown potential in distinguishing between credible and non-credible sources, but their ability to do so consistently and accurately remains challenging. The quality of the training data, the design of user experiences, and the implementation of tools to verify the credibility of outputs are all critical factors in enhancing the reliability of LLMs. The ability of LLMs to differentiate between primary and secondary sources is crucial, especially in fields such as law, finance, and healthcare, where the distinction determines the authority and application of information.

Primary sources are original materials that directly relate to an event, person, or subject of interest. They are considered authoritative and include legal documents such as constitutions, statutes, cases, and administrative rules. Secondary sources analyze, interpret, or summarize primary sources.

When citing sources, it is important to include both primary and secondary sources to provide a comprehensive view and support for the ideas presented. LLMs can assist in identifying when a primary source is cited by a secondary source, which is essential for proper attribution and understanding of the origin of ideas.

LLMs are limited by their capacity to discern the credibility of sources, and the reliability of the information they generate has been a concern. To mitigate this, it's crucial to ensure that LLM outputs are verifiable and traceable to credible sources.

- **Verifiability and User Experience**: Platforms like Perplexity.ai and ChatGPT are working towards providing users with citations and the ability to check the reliability of statements. A well-designed user experience (UX) can help users assess AI-generated content critically by providing transparency about the LLM's operations and information sources.
- **Dependence on Reliable Sources**: If an LLM is trained on reliable sources, it is considered reasonably trustworthy. For instance, a system trained on Wikipedia might be more reliable due to Wikipedia's robust mechanisms for sourcing credibility. However, some popular LLMs include sources designed to mislead, which raises concerns about their trustworthiness.
  - Cleaning and preparing large datasets for LLM training is essential.
  - Building models tailored to specific domains or tasks can improve efficiency and performance
  - Prioritizing high-quality data over sheer quantity can lead to more efficient learning
- **Real-Time Data and Mathematical Accuracy**: LLMs need help accessing real-time data, which can lead to outdated or irrelevant information being used in their outputs. They also struggle with mathematical calculations, as their responses are based on patterns in their training data rather than actual computation.

Challenges such as ensuring the accuracy and currency of information and managing potential biases highlight the need for human oversight and critical evaluation of LLM outputs.

Big Data, AI, Generative AI, and Large Language Models (LLMs) are interconnected technologies that collectively contribute to advancements in AI. Big Data provides the necessary input for training Gen AI models and LLMs. The vast amounts of data help these models learn complex patterns and improve their output quality.

AI, including Gen AI and LLMs, leverages Big Data to enhance analytics capabilities. It can analyze textual data, extract insights, and support decision-making processes.

In summary, Big Data fuels the development of AI systems by providing the necessary information for training. Gen AI uses this data to create new content across different media forms, while LLMs specialize in processing and generating human-like text. These technologies enhance data analytics and business intelligence by offering advanced capabilities for understanding and generating content.

REFERENCES

Bailly, A., Blanc, C., Francis, É., Guillotin, T., Jamal, F., Wakim, B. and Roy, P. 2022. Effects of dataset size and interactions on the prediction performance of logistic regression and deep learning models. *Computer Methods and Programs in Biomedicine* 213: 106504.

Bender, E. M., Gebru, T., McMillan-Major, A. and Shmitchell, S. 2021. On the dangers of stochastic parrots: can language models be too big? In *Proceedings of the 2021 ACM Conference on Fairness, Accountability, and Transparency (FAccT)* (pp. 610–623).

Brown, T. B., Mann, B., Ryder, N., Subbiah, M., Kaplan, J., Dhariwal, P., Neelakantan, A., Shyam, P., Sastry, G., Askell, A. and Agarwal, S. 2020. Language models are few-shot learners. *Advances in Neural Information Processing Systems* 33: 1877–1901.

Brown, T., Mann, B., Ryder, N., Subbiah, M., Kaplan, J.D., Dhariwal, P., Neelakantan, A., Shyam, P., Sastry, G., Askell, A. and Agarwal, S. 2020. Language models are few-shot learners. Arxiv preprint arXiv:2005.14165.

"Data Quality Considerations for AI Systems: Insights and Recommendations" by the World Economic Forum, highlights the importance of quality in AI and ML systems and its impact on model performance.

Devlin, J., Chang, M. W., Lee, K. and Toutanova, K. 2018. BERT: Pre-training of deep bidirectional transformers for language understanding. Arxiv preprint.

Halevy, P. N. and Pereira, F. 2009. The unreasonable effectiveness of data. *IEEE Intelligent Systems* 24(2): 8–12. doi: 10.1109/MIS.2009.36.

Mann, R.P., Mushtaq, F., White, A.D., Mata-Cervantes, G., Pike, T., Coker, D., Murdoch, S., Hiles, T., Smith, C., Berridge, D. and Hinchliffe, S. 2016. The problem with big data: Operating on smaller datasets to bridge the implementation gap. *Frontiers in Public Health* 4: 248.

Pawar, U., Beder, C., O'Reilly, R. and O'Shea, D. 2024. *On the Impact of Neighbourhood Sampling to Satisfy Sufficiency and Necessity Criteria in Explainable AI.* CLEaR.

Sapir, M. 2018. Learnable: Theory vs applications. ArXiv. /abs/1807.10681.

Sculley, D., Holt, G., Golovin, D., Davydov, E., Phillips, T., Ebner, D., Chaudhary, V., Young, M., Crespo, J.F. and Dennison, D. 2015. Hidden technical debt in machine learning systems. *Advances in Neural Information Processing Systems.*

Sun, C., Shrivastava, A., Singh, S. and Gupta, A. 2017. Revisiting unreasonable effectiveness of data in deep learning era. In *Proceedings of the IEEE International Conference on Computer Vision* (pp. 843–852).

Ulrich, H., Kock-Schoppenhauer, A. K., Dep.penwiese, N., Gött, R., Kern, J., Lablans, M., Majeed, R. W., Stöhr, M. R., Stausberg, J., Varghese, J., Dugas, M. and Ingenerf, J. 2022. Understanding the nature of metadata: Systematic review. *Journal of Medical Internet Research* 24(1): e25440.

Understanding Data Quality in Big Data by Zillner, S., Mikula, M. and Härting, R. 2018. Big data quality: Data-centric quality management in data-driven enterprises. In *Smart City Symposium.*

10 Aligning Big Data and AI Strategy with Business Goals

AI has emerged as a transformative force across industries in today's rapidly evolving technological landscape. A Big Data/AI strategy is a structured, goal-oriented approach for businesses aiming to leverage AI effectively. A well-defined AI strategy is not just an option; it's a necessity (Figure 10.1).

The strategy helps align initiatives with an organization's objectives, ensuring AI investments drive measurable business outcomes, such as cost reduction, revenue growth, or improved customer experience. The drivers for an AI strategy are:

1. **Navigating Complexity**: AI technologies are diverse and complex, encompassing machine learning, natural language processing, and computer vision. An AI strategy helps organizations navigate this complexity by providing a clear technology selection, implementation, and integration roadmap.
2. **Maximizing Value**: Without a strategic approach, investments in AI can lead to suboptimal outcomes. An AI strategy ensures that resources are allocated efficiently, projects are aligned with business objectives, and the full potential of AI technologies is realized.
3. **Data Utilization**: AI relies heavily on quality data. An AI strategy emphasizes data governance, ensuring the organization effectively manages, cleans, and leverages its data assets to fuel AI systems.
4. **Competitive Advantage**: Organizations can differentiate themselves in the market by identifying areas where AI can create efficiencies or offer

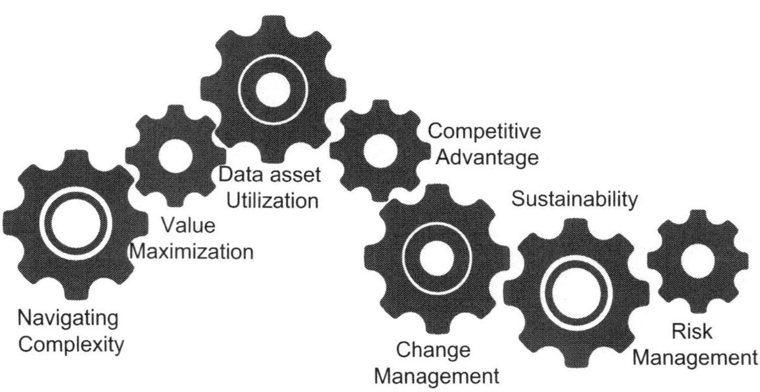

FIGURE 10.1 Drivers for AI Strategy.

DOI: 10.1201/9781003190325-10

unique capabilities. It enables businesses to innovate, improve operational efficiency, and enhance customer experiences, leading to a stronger market position.
5. **Sustainability**: AI strategy ensures that AI initiatives are scalable across the organization and adaptable to future needs. This includes the proper infrastructure, tools, and processes to manage the AI lifecycle.
6. **Change Management**: AI transformation requires adapting the organization's culture, workforce, and processes. A strategic approach helps manage this change, ensuring employees are upskilled and AI technologies are smoothly integrated into existing operations.
7. **Managing Risks**: Implementing AI involves risks related to data privacy, ethical considerations, and regulatory compliance. An AI strategy addresses these risks proactively, ensuring the responsible and compliant use of AI technologies.

Establishing a good AI strategy can help organizations harness AI's full potential, avoid pitfalls, and create a clear roadmap for long-term success.

10.1 HOW TO DEVELOP AN EFFECTIVE AI STRATEGY

AI projects may become disjointed or fail due to a lack of resources, expertise, or direction. A well-defined AI strategy helps prioritize high-impact use cases, optimize budgets, and allocate the right talent. The following aspects need to be taken into consideration when developing a good AI strategy:

1. **Align with Business Objectives**: The first step in developing an AI strategy is ensuring it aligns with the broader business strategy. This involves understanding the organization's goals, challenges, and opportunities, and identifying how AI can support these objectives. For instance, if a business aims to enhance customer satisfaction, AI-powered chatbots or personalized recommendations can be integrated to achieve this goal.
2. **Assess Current Capabilities/Readiness**: Evaluate the organization's existing technological infrastructure, data quality, and human resources. This assessment helps identify gaps and determine readiness for AI adoption. It also highlights the need to upskill employees or acquire new talent.
3. **Define Use Cases and Prioritize**: Identify specific use cases where AI can deliver the most value. Prioritize these use cases based on potential impact, feasibility, and alignment with strategic goals. This step involves developing a proof of concept (PoC) to validate the feasibility and value of selected use cases.
4. **Develop a Roadmap**: Create a detailed AI roadmap outlining the implementation phases, resource allocation, and key milestones. The roadmap should include plans for technology selection, integration, and scaling of AI solutions. It should also address change management and training needs.
5. **Establish Governance and Ethics**: Implement governance structures to oversee AI initiatives, ensuring they adhere to ethical standards and

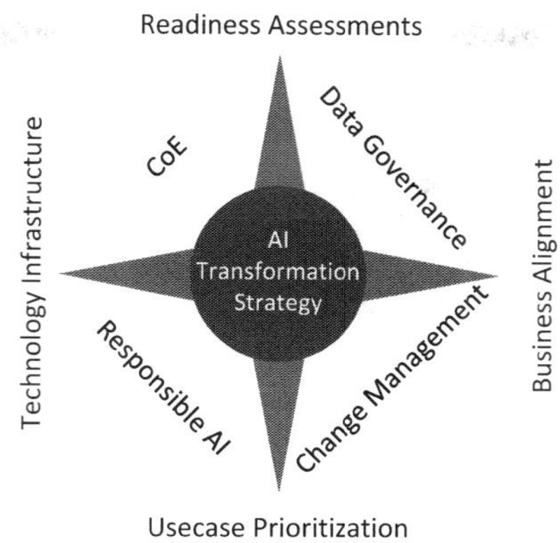

FIGURE 10.2 Key components of an AI Strategy.

regulatory requirements. Establishing clear guidelines for data privacy, transparency, and bias mitigation is crucial for responsible AI use.

6. **Monitor and Iterate**: An AI strategy should be dynamic, with mechanisms for monitoring progress and iterating based on feedback and changing business needs. Regular reviews and adjustments ensure that the strategy remains aligned with evolving goals and technological advancements (Figure 10.2).

10.1.1 Align with Business Objectives

Ensure that AI initiatives are directly linked to business goals such as growth, efficiency, or innovation. This alignment helps prioritize projects that offer the most significant strategic benefit.

- **Integrated Planning**: Incorporate AI strategy into the overall business planning process. This involves collaboration between AI specialists and business leaders to ensure that AI projects support broader business objectives and deliver measurable results.
- **Value-Driven Implementation**: Focus on delivering tangible value from AI investments. This includes setting clear performance metrics and KPIs that align with business outcomes, such as increased revenue, cost savings, or enhanced customer satisfaction.
- **Cross-Functional Collaboration**: Foster collaboration between departments—such as IT, operations, and marketing—to ensure that AI solutions are effectively integrated into business processes and strategies.

By developing a comprehensive AI strategy that aligns with business goals, organizations can harness AI's transformative potential while mitigating risks and ensuring sustainable success.

10.1.2 Assess Current Capabilities

Readiness assessment frameworks provide structured approaches to evaluate an organization's preparedness for AI and data-driven initiatives. Analytical Maturity and Readiness Assessments are critical frameworks and methods used to evaluate how well-prepared an organization is to adopt, implement, and scale data-driven initiatives, including AI, analytics, and data science. These assessments help identify strengths, gaps, and areas for improvement in data and analytics capabilities, providing a clear roadmap for enhancing maturity and readiness. A typical readiness assessment framework includes several key components:

a. **Data Readiness**: This component evaluates the quality, accessibility, and management of data within the organization. It includes assessing data sources, integration processes, governance practices, and security measures. High-quality, well-governed data is essential for effective AI and analytics, and this component helps identify any deficiencies in data management that need to be addressed.
b. **Analytical Maturity**: Assessing analytical maturity involves evaluating an organization's capabilities and readiness to effectively leverage data analytics and AI technologies. This assessment provides insights into the strengths and weaknesses of an organization's analytical processes, tools, and culture, helping identify areas for improvement and guiding AI transformation efforts.
c. **Technical Readiness**: This component assesses the organization's technical infrastructure and capabilities. It involves evaluating the existing technology stack, including hardware, software, and analytics tools. Organizations should assess whether their technology infrastructure can support the requirements of AI and data analytics, such as scalability, processing power, and integration capabilities.
d. **People/Skill Readiness**: This component examines the skills and expertise of the organization's workforce. It includes assessing employees' proficiency in areas such as data analysis, machine learning, and AI technologies. Identifying skill gaps and training needs helps prepare the workforce for AI initiatives and ensures they have the necessary competencies to support successful implementation.
e. **Process Readiness**: This component evaluates the organization's existing processes and workflows related to data and analytics. It includes assessing the maturity of analytics processes, decision-making frameworks, and the integration of data insights into business operations. Ensuring processes are well-defined and aligned with AI goals is critical for successful outcomes.
f. **Cultural Readiness**: This component assesses the organization's culture and attitude towards data-driven decision-making and AI. It involves evaluating whether there is a strong data-driven mindset, support for innovation, and openness to change. A positive and supportive culture is essential for fostering successful AI adoption and ensuring that the organization embraces data-driven strategies.

Identifying strengths, weaknesses, and areas for improvement creates a canvas for organizations based on which targeted strategies and roadmaps can be developed.

The following methods provide an approach to assessing an organization's readiness for analytics and AI initiatives. They help develop clear roadmaps to improve analytical maturity and capitalize on data-driven opportunities.

1. **Surveys and Questionnaires**: Distribute surveys across departments to gather input on how data and analytics are used and what challenges exist.
2. **Workshops and Interviews**: Conduct workshops and in-depth interviews with key stakeholders (e.g., C-suite, business unit leaders, data scientists) to gather insights on analytical capabilities and readiness.
3. **Benchmarking**: Use external benchmarks to compare the organization's maturity and readiness with industry peers. Look for studies or industry reports on data and AI maturity (e.g., McKinsey, Gartner).
4. **Gap Analysis**: After conducting the assessment, perform a gap analysis to identify areas where the organization is underperforming. This helps prioritize where resources should be focused.
5. **SWOT Analysis**: Apply SWOT to assess the strengths, weaknesses, opportunities, and threats related to analytics capabilities.
6. **Data and Technology Audits**: Perform a technical audit of the data infrastructure, tools, and platforms to assess their adequacy for supporting advanced analytics and AI.
7. **Stakeholder Engagement**: Engage stakeholders across business, IT, and analytics teams to ensure the assessment captures technical and business perspectives.

Below are key frameworks and methods for assessing analytical maturity and readiness, along with how to conduct these assessments:

<table>
<tr><th colspan="2">Deloitte's Analytical Maturity Model</th></tr>
<tr><td>Levels of Maturity:
1. Basic (Ad-hoc): Analytics efforts are scattered and isolated. Data analysis is ad hoc, with no formal processes or tools.
2. Localized (Repeatable): Certain departments or functions have developed analytics capabilities, but an enterprise-wide strategy is needed.
3. Integrated (Defined): Analytics is integrated across multiple business units with standardized tools and processes.
4. Advanced (Managed): Data-driven decision-making is a key driver of business strategy, and predictive or prescriptive analytics are common.
5. Transformational (Optimized): Analytics is fully embedded in the organization with AI/ML initiatives at scale.</td><td>Steps:
1. Identify key business processes and functions that rely on data analytics (e.g., marketing, operations, finance).
2. Evaluate each process against the five maturity levels, asking questions such as:
• Do we use analytics in this area? How standardized is it?
• Are decisions based on data-driven insights?
• Are advanced techniques like machine learning or AI being used?
3. Benchmark each department/process's maturity level to identify which ones are lagging and need attention.
4. Develop a roadmap to move departments from lower to higher maturity stages, focusing on scaling best practices across the organization.</td></tr>
</table>

(Continued)

<table>
<tr><th colspan="2">Gartner's Analytics Maturity Model</th></tr>
<tr><td>Stages of Maturity:
Unaware: No formal analytics capabilities.
Opportunistic: Some departments or teams experiment with analytics on an ad hoc basis.
Standards: Analytics efforts are standardized across departments.
Enterprise: Analytics are integrated into enterprise processes and systems.
Advanced: Predictive analytics, ML, and advanced modeling are widely used.
Transformative: Analytics and AI are fully embedded in decision-making processes across the organization.</td><td>How to Perform the Assessment:
1. Break down the organization into functions and processes that rely on analytics (e.g., supply chain, customer service, product development).
2. For each function, evaluate where it sits on Gartner's six-stage maturity curve by asking:
• Are data-driven insights used consistently for decision-making?
• How sophisticated are the analytics techniques being used?
3. Are advanced tools (e.g., machine learning models) part of the analytics toolkit?
4. Based on the findings, identify key gaps. For example, marketing uses advanced analytics while operations are still at a basic level.
5. Prioritize areas for improvement and create a maturity roadmap.</td></tr>
<tr><th colspan="2">TDWI Analytics Maturity Model
Five dimensions: organization, infrastructure, data management, analytics, and governance</th></tr>
<tr><td>Stages of Maturity:
Nascent: Limited analytics, basic reporting, and siloed data efforts.
Pre-Adoption: Some departments are using analytics, but not systematically.
Early Adoption: Organization-wide awareness of analytics' value and tools/processes will start becoming standardized.
Corporate Adoption: Widespread use of analytics across the business, with advanced capabilities (e.g., predictive analytics) being adopted.
Mature/Visionary: Data-driven culture is fully embedded, and AI/ML initiatives drive competitive advantage.</td><td>How to Perform the Assessment:
1. Evaluate each of the five key dimensions (organization, infrastructure, data management, analytics, and governance) by asking questions like:
• Is there executive sponsorship and a clear vision for analytics?
• How mature is the data infrastructure (e.g., data lakes, DW)?
• How advanced are the analytics tools and models (e.g., predictive, prescriptive analytics)?
• Are data governance and security policies in place?
2. Assign a maturity score (from 1 to 5) to each dimension.
3. Plot the scores on a radar chart or matrix to identify areas of strength and weakness.
4. Develop targeted strategies to improve lower-scoring areas.</td></tr>
</table>

(Continued)

AI Maturity Framework (McKinsey)

Key dimensions: strategy, technology, people, data, and governance.

Dimensions Assessed:

Strategy: How aligned is the AI strategy with overall business goals?

Technology: Are the tools, platforms, and AI models in place to support advanced analytics?

People: Do you have the right talent (data scientists, AI engineers) to drive initiatives?

Data: Is the data infrastructure ready to support AI and analytics?

Governance: Are data security, privacy, and ethical AI frameworks in place?

How to Perform the Assessment:

1. Evaluate each of the five key dimensions by conducting surveys or interviews with key stakeholders across the organization.
2. Score each dimension from 1 (low) to 5 (high) to assess readiness for AI and advanced analytics.
3. Use the assessment to create a prioritized action plan that includes investments in technology, hiring of skilled personnel, improving data quality, or revisiting the AI strategy.

Data Maturity Model by DAMA-DMBOK (Data Management Body of Knowledge)

Dimensions:

Data Governance: How effectively does the organization manage and control its data?

Data Architecture: How robust and scalable is the underlying data infrastructure?

Data Quality: Is the data fit for purpose, clean, and reliable?

Data Integration: How well integrated is data across various functions and systems?

Metadata Management: Are data definitions and lineage well-documented?

How to Perform the Assessment:

1. Break down the organization's data management practices by each dimension (e.g., governance, architecture, integration).
2. Rate each dimension from 1 to 5 (1 = Basic, 5 = Optimized).
3 Identify gaps that hinder effective analytics or AI initiatives, such as poor data quality or lack of governance.
4. Develop a data improvement roadmap that focuses on areas with low maturity (e.g., setting up a data governance framework).

Capability Maturity Model Integration (CMMI) Data Management Maturity (DMM)

Maturity Levels:

Initial: Data management processes are ad hoc and reactive.

Managed: Basic data management practices are in place.

Defined: Data management practices are standardized across the organization.

Quantitatively Managed: Data processes are measured and controlled using metrics.

Optimizing: Data management processes are continuously improving.

How to Perform the Assessment:

Step 1: Assess data management capabilities across dimensions such as data governance, data architecture, and data quality management.

Step 2: Evaluate each dimension using CMMI's five maturity levels, assigning scores based on how well-established each practice is.

Step 3: Prioritize gaps and identify areas where the organization can improve data management to support analytics and AI initiatives.

Step 4: Develop a phased plan to progress from lower to higher maturity levels.

10.1.3 Define Use Cases and Prioritize

Identifying suitable Use Cases and pilot projects for AI implementation involves establishing clear criteria to ensure that selected projects are aligned with organizational goals and have the potential to demonstrate value. Effective selection criteria help prioritize projects that offer the greatest impact while managing risk and resource allocation.

- **Strategic Alignment**: Pilots should be chosen based on how well they align with the organization's strategic objectives and business priorities. Projects that address critical pain points or have the potential to drive significant improvements in key areas such as customer experience, operational efficiency, or revenue generation should be prioritized. A Proof of Concept (PoC) is an important prerequisite for a Pilot. A small PoC is a great idea for testing the feasibility of a pilot.
- **Feasibility**: This includes evaluating the technical feasibility of the project, such as the availability of data, the complexity of the required AI models, and the integration with existing systems. Assessing the availability of skilled personnel as well as the estimated time and cost required for implementation is also crucial.
- **Impact Potential**: Identifying projects with clear, quantifiable goals and metrics for success (such as cost savings, increased productivity, potential revenue, customer impact, or enhanced decision-making) helps in demonstrating the value of AI and building a business case for broader adoption. Time to value: Prioritize AI projects based on how quickly they can deliver measurable business value, such as automating manual tasks or improving forecasting accuracy.
- **Risk**: Choosing projects with manageable risks (such as technical, data privacy concerns, ethical implications, regulatory issues, or potential disruptions to existing processes) and a clear plan for mitigation can help avoid unforeseen challenges and ensure successful implementation.

Use cases, PoCs, pilots, minimum viable products (MVPs), projects, and initiatives that have strong support from key stakeholders and align with their interests and needs are more likely to succeed. Engaging stakeholders early in the process and incorporating their feedback helps gain buy-in and facilitates smoother project execution (Figure 10.3).

Identifying important use cases, pilots, or projects for AI initiatives requires structured frameworks and methods to ensure alignment with strategic business goals, feasibility, and long-term value. Here are some frameworks and techniques to help organizations of all sizes prioritize and create a roadmap for AI initiatives:

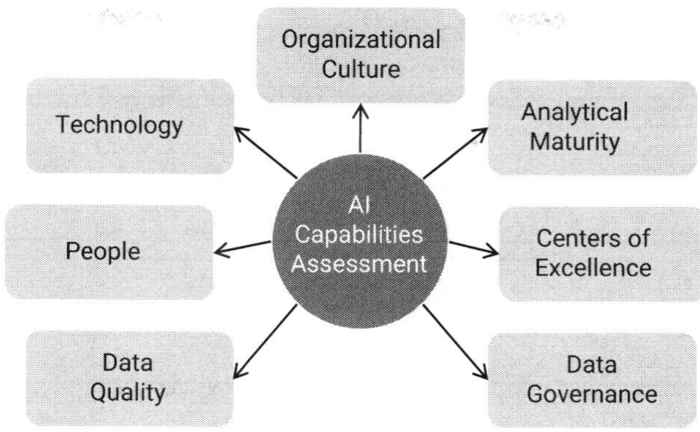

FIGURE 10.3 Drivers for prioritizing Usecase, PoC or MVP selection.

ICE Scoring Model (Impact, Confidence, Ease)

Prioritizes AI projects based on their impact, confidence in success, and ease of implementation.

Application:

- **Impact**: How significantly will the AI use case affect business outcomes (revenue, cost savings, customer satisfaction)?
- **Confidence**: Based on available data, expertise, and technology maturity, how confident are you in the AI solution's success?
- **Ease**: How easy or challenging is implementing the AI project, considering resources, technology, and organizational readiness?

Best for: Small and medium-sized organizations to quickly prioritize use cases for maximum return with minimal risk.

McKinsey's Three Horizons of Growth Framework

Identifies AI use cases and projects across three horizons—short-term, mid-term, and long-term.

Application:

- **Horizon 1 (Core AI)**: Immediate AI initiatives that optimize existing operations (e.g., process automation).
- **Horizon 2 (Adjacency AI)**: Medium-term AI projects that explore adjacent opportunities (e.g., expanding into new markets and services).
- **Horizon 3 (Transformational AI)**: Long-term projects with disruptive AI technologies, reshaping the business model (e.g., autonomous systems, deep personalization).

Best for Large organizations to balance short-term gains with long-term innovation while building AI capability incrementally.

AI Canvas (AI Use Case Canvas)

A framework similar to the Business Model Canvas that helps systematically define and evaluate AI use cases.

Components:

- **Problem**: What business problem are you solving with AI?
- **Data**: What data is available? What data quality is required for the use case?
- **Models**: What AI/ML models will solve the problem?
- **Evaluation**: How will you measure success (KPIs, accuracy, impact)?
- **Risks**: What risks (technical, operational, ethical) are involved in this use case?

Best for: Any size of an organization looking for a structured, step-by-step approach to evaluating and prioritizing AI use cases.

Opportunity-Solution Mapping

Identifies key AI opportunities by breaking them down into specific business needs and solutions.

Application:

- **Step 1**: Map out key business opportunities (e.g., improving customer experience and reducing operational costs).
- **Step 2**: Brainstorm potential AI-driven solutions for each opportunity (e.g., customer service chatbot, predictive maintenance).
- **Step 3**: Evaluate the feasibility and impact of these solutions.

Best for: Small and medium organizations looking for quick ways to identify valuable use cases that align with business needs.

Value vs. Complexity Matrix (Quick Wins Framework)

Helps prioritize AI use cases by plotting them on a 2x2 matrix of business impact and ease of implementation

Application:

- **Quick Wins**: High value, low complexity—prioritize these for pilots.
- **Strategic Initiatives**: High value, high complexity—plan for long-term AI roadmap.
- **Low-Hanging Fruit**: Low value, low complexity—consider only if resources allow.
- **Avoid**: Low value, high complexity—deprioritize or discard.

Best for: Any size organization to visually assess and sequence AI use cases.

CRISP-DM (Cross-Industry Standard Process for Data Mining)

A project management methodology for AI initiatives, ensuring systematic execution from business understanding to deployment.

Stages:

- **Business Understanding**: Identify AI use cases based on specific business goals.
- **Data Understanding**: Assess data availability and quality for each AI project.
- **Modeling**: Develop AI models for each use case.
- **Evaluation**: Test models and measure business impact.
- **Deployment**: Deploy AI models in production.

Best for: Medium and large organizations for creating repeatable AI processes and piloting initiatives.

OKR (Objectives and Key Results) Framework

Aligns AI initiatives with business objectives by focusing on measurable outcomes.

Application:

- **Step 1**: Define business objectives (e.g., improve customer retention by 10%).
- **Step 2**: Identify key results that AI initiatives can deliver (e.g., AI-based churn prediction model).
- **Step 3**: Track progress and update based on measurable results.

Best for: Medium to large organizations to align AI initiatives with strategic business objectives and outcomes.

Lean AI Framework

Focuses on building AI initiatives in an agile, iterative way by experimenting, learning, and scaling.

Application:

- **Step 1**: Start with small experiments or pilots, testing AI use cases in real-world conditions.
- **Step 2**: Learn from results, iterate, and improve the AI models.
- **Step 3**: Scale up successful AI initiatives across the organization.

Best for Startups or smaller organizations that want a low-risk, iterative approach to developing AI projects.

Business Model Innovation Framework

Guides organizations in using AI to innovate their business models and explore new opportunities.

Application:

- **Step 1**: Identify AI use cases that can add value to current business models (e.g., automating customer service or creating new data-driven services).
- **Step 2**: Evaluate how AI can open up new revenue streams, disrupt existing models, or create a competitive advantage.
- **Step 3**: Pilot these AI-driven business models to test feasibility.

Best for: Large organizations exploring transformative AI use cases or new business models.

AI Maturity Model (AIMM)

Evaluates an organization's AI readiness across multiple dimensions—strategy, culture, people, data, technology, and governance—and help prioritize AI projects based on the current level of maturity.

Stages:

- **Ad-hoc**: AI initiatives are uncoordinated and experimental.
- **Repeatable**: AI use cases are more consistent but still isolated.
- **Defined**: AI initiatives are embedded within business processes.
- **Managed**: AI is integrated with defined metrics and governance.
- **Optimized**: AI is fully aligned with the business strategy and is continuously improving.

Best for: Medium and large organizations to assess AI maturity and develop a roadmap that includes pilots and scalable initiatives.

Each framework and method outlined above offers a structured approach to identifying, prioritizing, and creating a roadmap for AI initiatives, ensuring they are impactful, scalable, and aligned with business objectives.

10.2 CHANGE MANAGEMENT IN AI TRANSFORMATIONS

Change management plays a crucial role in AI transformations because it helps organizations navigate the complex shifts in processes, technology, and people that are essential for successful AI adoption. AI transformations often bring about significant changes in workflows, decision-making, and organizational culture, and effective change management ensures that these changes are accepted, understood, and supported by all stakeholders.

Key elements of Change Management in AI Transformations:

- **Addressing Resistance to Change**: AI can sometimes lead to fear or uncertainty among employees, particularly around job security or new skill requirements. A structured change management plan helps mitigate resistance by clearly communicating the benefits of AI and involving employees in the transition process.
- **Training and Upskilling**: AI transformations often require employees to adopt new tools and technologies. Change management helps identify skill gaps and ensures that training and upskilling programs are in place so employees can adapt to the new systems.
- **Aligning with Business Objectives**: Change management ensures that AI initiatives align with the organization's overall goals and are not perceived as disruptive. It bridges the gap between technical innovation and business strategy by fostering collaboration between leadership, technology teams, and end users.
- **Cultural Change and Mindset Shift**: AI adoption involves a shift from traditional processes to data-driven decision-making. Change management helps foster a culture that embraces innovation, experimentation, and continuous learning.
- **Communication and Stakeholder Engagement**: Successful AI transformations rely on clear and consistent communication with stakeholders about AI initiatives' purpose, progress, and expected outcomes. Change management frameworks provide tools to engage stakeholders early, ensuring buy-in from key decision-makers and end users.
- **Managing Organizational Impacts**: AI-driven changes can impact various aspects of the business, including roles and responsibilities, workflows, and customer interactions. Change management frameworks help map out these impacts and develop strategies to manage them.
- **Ensuring Long-Term Success**: Beyond the initial implementation, change management supports the sustained adoption of AI by continuously monitoring and adjusting processes, ensuring that the changes are deeply embedded in the organization's operations.

10.2.1 Frameworks for AI Change Management

- **Prosci ADKAR Model**: The ADKAR (Awareness, Desire, Knowledge, Ability, Reinforcement) model is widely used for managing organizational change. It emphasizes building awareness of the need for change, creating a desire among employees to participate, providing knowledge and training, ensuring the ability to implement the change, and reinforcing the change to make it stick. The ADKAR model can guide organizations in preparing employees for the shifts brought by AI, ensuring they understand the "why" behind AI and are equipped with the necessary skills to embrace new technologies.

- **Kotter's 8-Step Change Model**: This model focuses on creating urgency, building a guiding coalition, developing a vision, and empowering action. The steps include establishing urgency, forming a powerful coalition, creating a vision for change, communicating the vision, removing obstacles, creating short-term wins, building on the change, and embedding new approaches in the culture. For AI transformation, this model ensures that leadership is actively involved, a clear vision for AI initiatives is communicated, and quick wins (e.g., successful pilots) are showcased to generate momentum.
- **Lewin's Change Management Model**: The model consists of three stages: Unfreeze, Change, and Refreeze. It can be used to shift from legacy systems to AI technologies, with an emphasis on making the transition permanent by embedding AI tools and processes into everyday workflows.
- **McKinsey 7-S Framework**: The 7-S framework focuses on aligning seven internal elements of an organization: Strategy, Structure, Systems, Shared Values, Style, Staff, and Skills. This framework ensures that AI adoption isn't just a technological change but also considers structural, cultural, and skill-based impacts, aligning them with the organization's strategy.
- **The SCARF Model (David Rock)**: The SCARF model focuses on human factors in change management: Status, Certainty, Autonomy, Relatedness, and Fairness. AI implementations can disrupt traditional job roles and workflows, leading to stress and uncertainty. The SCARF model addresses these human factors, ensuring that AI adoption is approached in a way that minimizes psychological discomfort.
- **Agile Change Management**: This framework integrates agile methodologies into change management, focusing on iterative change, continuous feedback, and flexibility. Since AI projects often evolve rapidly, agile change management enables organizations to adapt to new developments in AI tools and models while ensuring that employees and processes can keep up with these changes.
- **The People-Centered Implementation (PCI) Methodology**: PCI focuses on engaging people throughout the change process and ensuring their needs are met. AI transformations can intimidate staff due to concerns over automation and job security. PCI helps organizations address these concerns and foster collaboration.

By applying the appropriate frameworks, organizations can ensure that their AI transformations are technologically successful and widely accepted and integrated by their workforce.

10.2.2 Skill Assessment and Talent Development

Skill assessment and gap analysis are crucial for aligning an organization's human resources (HR) with its AI transformation goals. As AI technologies become integral to business operations, understanding employees' current skill sets and identifying gaps is essential for effectively implementing and utilizing AI solutions.

Several frameworks and approaches can be used to assess and develop AI-related skills as part of an organization's AI strategy and transformation programs. These frameworks ensure that AI capabilities are aligned with business needs while addressing gaps in skills, fostering continuous development, and promoting organizational readiness for AI initiatives.

- **AI Capability Maturity Model**: This model assesses an organization's AI maturity in various dimensions, including technical capabilities, data management, culture, and talent. It helps identify current skill levels and areas where development is needed, such as data science, machine learning engineering, AI ethics, and business translation. Organizations are categorized from basic awareness to advanced AI implementation, with corresponding skill requirements. It's a helpful framework for evaluating readiness, setting benchmarks, and identifying upskilling or hiring needs.
- **AI Competency Framework (World Economic Forum)**: This framework provides a structured approach to identifying essential AI competencies across various job roles and functions and provides clear guidance for organizations looking to develop role-specific AI competencies. It divides competencies into four areas:
  - **Technical Skills**: Machine learning, data engineering, natural language processing, etc.
  - **Ethical Skills**: Understanding bias, privacy concerns, and responsible AI development.
  - **Business Skills**: Translating AI into business impact, identifying AI opportunities, and aligning with strategy.
  - **Leadership Skills**: AI project management, governance, and ethical leadership.
- **European AI Skills Framework (AI4EU)**: This framework is focused on mapping the AI skills required in European industries and across various job profiles. It identifies skills in areas such as AI development, data science, ethics, and business process integration, and includes soft skills like problem-solving and communication. Organizations can use it to assess skills gaps, especially when considering regulatory and ethical standards for AI development and deployment.
- **The AI Skills Taxonomy (OECD)** categorizes AI-related skills based on industry demands and academic research. It's designed to aid both the assessment and development of AI skills. The taxonomy can guide training programs and help assess skills gaps across technical and non-technical employees. The taxonomy divides skills into:
  - **Foundational AI Skills**: Basic AI literacy, understanding AI's impact on business, and ethics.

  - **Technical AI Skills**: Programming, data analytics, deep learning, model building.
  - **Operational AI Skills**: Deployment, scaling, monitoring, MLOps.
- **Digital Capability Framework**: Though broader in scope, this framework includes AI as part of digital transformation initiatives and can be tailored to focus specifically on AI capabilities. It covers technical (AI/ML, data science) and organizational (change management, business innovation) capabilities. It helps organizations design programs for continuous upskilling and AI integration by mapping current digital and AI skills against future needs.
- **T-Shape Skills Development Framework**: This framework focuses on developing "T-shaped" skills, where individuals have a breadth of understanding across AI-related areas (the horizontal bar) and depth in a particular specialization (the vertical bar). It encourages employees to build foundational knowledge across data science, AI ethics, and business integration while developing deep expertise in natural language processing, computer vision, or AI strategy. It is ideal for upskilling programs, ensuring employees are versatile enough to contribute across multiple AI projects while specializing in key areas of expertise.
- **AI Competency Framework for Leadership (Harvard Business Review)**: This framework focuses on upskilling leaders and managers to make informed decisions on AI projects, understand AI's potential, and manage AI-driven transformations. It encourages leadership skills in AI governance, ethical AI, change management, and strategic planning. It helps leadership teams build an AI-literate organization by ensuring they can effectively guide AI initiatives.

Organizations can develop a custom AI skills matrix based on their specific needs and AI roadmap. A well-tailored matrix may categorize skills by technical, analytical, ethical, and strategic roles, helping organizations plan for targeted upskilling or hiring. It may offer flexibility to address specific skills gaps across functions, ensuring employees develop the necessary competencies aligned with the organization's AI goals.

10.3 CENTERS OF EXCELLENCE (COES) IN AI TRANSFORMATIONS

CoEs are pivotal in an organization's AI strategy by serving as centralized hubs for best practices, governance, innovation, and capability building. They help streamline and optimize the deployment of AI across business units, ensuring alignment with the overall strategy and scaling AI efforts systematically.

In an AI strategy, CoEs are the cornerstone for driving innovation, scaling AI solutions, enforcing governance, and aligning AI initiatives with business objectives. They ensure that AI efforts are consistent, scalable, and effectively governed, leading to greater organizational impact and efficiency. By focusing on building expertise, fostering collaboration, and maintaining oversight, AI CoEs help organizations unlock the full potential of AI technologies.

CoEs are organizational structures or dedicated teams that focus on developing, promoting, and standardizing best practices, innovation, and efficiency in specific domains, such as technology, analytics, or artificial intelligence (AI). CoEs serve as hubs of expertise within an organization, aimed at driving excellence by centralizing specialized knowledge, resources, tools, and processes that can be leveraged across various business units (Figure 10.4).

Here's how CoEs fit into AI strategy:

- **Driving AI Innovation and Experimentation**: CoEs foster innovation by providing a dedicated space for experimenting with new AI models, tools, and technologies. They incubate new AI solutions, allowing pilot projects to be tested in a controlled environment before scaling them across the organization. AI CoEs work closely with business units to identify potential AI use cases and are often responsible for developing proof-of-concept initiatives to demonstrate the value of AI projects.
- **Standardization and Governance**: AI CoEs ensure standardization across AI projects, including the use of common frameworks, tools, and methodologies (e.g., ML/LLMOps, responsible AI practices). They develop and enforce governance policies related to data ethics, privacy, fairness, explainability, and regulatory compliances. AI CoEs play a key role in creating AI governance frameworks to manage the risks associated with AI, such as bias and model drift, ensuring AI systems remain auditable and transparent.
- **Scaling AI Initiatives**: A CoE serves as a central body to scale successful AI projects from pilot phases to full-scale implementations across different business units or geographies. They support AI adoption by providing frameworks for AI deployment, ensuring that projects are production-ready and aligned with business objectives. Centralizing resources enables faster, more efficient scaling of AI models and reduces duplication of efforts across departments.

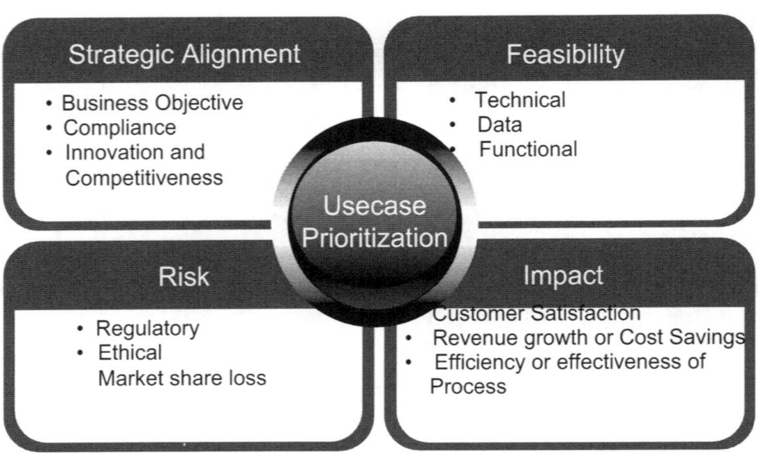

FIGURE 10.4 Core Responsibilities of an AI CoE.

- **Building AI Expertise and Talent Development**: CoEs are critical in developing internal AI talent by offering training, mentorship, and resources for upskilling employees in areas like data science, machine learning, and AI technologies. They create learning paths to help staff become proficient in AI tools, methodologies, and ethical practices. Additionally, AI CoEs serve as a center for knowledge sharing, disseminating lessons learned from pilot projects and case studies, thereby creating a shared knowledge base within the organization.
- **Technology Selection and Infrastructure**: AI CoEs play a key role in technology selection by helping the organization in evaluating and choosing the right AI platforms, tools, and infrastructure. They establish a comprehensive tech stack for AI, ensuring that components like data ingestion, model development, monitoring, and management tools are standardized and scalable. CoEs work closely with IT and data teams to ensure that the underlying data infrastructure (e.g., data lakes, cloud platforms) is robust enough to efficiently support AI workloads.
- **Alignment with Business Strategy**: CoEs ensure that AI initiatives are closely aligned with the overall business strategy by collaborating with senior leadership to identify key areas where AI can drive value (e.g., cost reduction, revenue generation, customer experience). They help prioritize AI projects by assessing their potential impact on business objectives, ensuring that resources are allocated to the most strategically important projects. Additionally, CoEs are responsible for measuring the success of AI initiatives using KPIs, such as ROI, productivity gains, and customer satisfaction improvements, ensuring that AI projects contribute directly to the organization's bottom line.
- **Ethical and Responsible AI Development**: AI CoEs are responsible for developing and enforcing ethical AI practices, ensuring that models are designed with fairness, transparency, and bias mitigation in mind. They guide responsible AI development across areas such as explainability, algorithmic accountability, and ensuring that AI outputs align with societal norms and regulatory requirements. CoEs also ensure compliance with data privacy laws (e.g., GDPR, CCPA) and industry-specific regulations.
- **Cross-functional collaboration and Culture Building**: AI CoEs promote a data-driven culture throughout the organization by encouraging collaboration between data scientists, engineers, business units, and executives. They serve as a bridge between technical and non-technical teams, ensuring AI strategies are understood and supported at all levels. CoEs also play a key role in embedding AI into everyday processes by promoting a mindset of continuous learning and experimentation, fostering AI adoption at the grassroots level.
- **Cost Efficiency and Resource Optimization**: By centralizing AI expertise, CoEs enable resource sharing across projects, ensuring that the organization can save money in duplicating efforts across different departments. CoEs track the ROI of AI initiatives, ensuring that projects remain within budget and deliver measurable business outcomes.

- **AI Strategy and Long-Term Vision**: CoEs help the organization in defining and continuously refining its AI strategy, ensuring it evolves as the company grows and new technologies emerge. They keep the organization updated on AI trends and developments, advising leadership on future AI investments. The CoE ensures that AI remains a strategic priority and that the organization can adapt to changes in the competitive landscape.

While CoEs are often associated with large organizations, they are applicable to businesses of all sizes—small, medium, and large. The structure and scope of a CoE can be scaled to align with the organization's size, resources, and goals.

10.4 OTHER CONSIDERATIONS IN AI TRANSFORMATIONS

In addition to the AI strategy roadmap, the following best practices, execution framework, and essential enabling functions must be established:

- **Data Management and AI Governance**: Establish robust data management practices, including data quality, integration, and governance policies. Ensure compliance with ethical standards and privacy regulations to address the ethical, operational, and regulatory aspects of AI implementation. This ensures proper oversight and alignment with organizational values.
- **Technology Infrastructure and Technology Selection**: Plan for the necessary technology infrastructure to support AI initiatives, including hardware, software, and cloud computing resources. This ensures scalability and adaptability as technology evolves.
  - **Cloud & AI Infrastructure**: As AI initiatives grow, leverage scalable and flexible AI platforms and cloud-based solutions to support increasing data and computational needs.
  - **Adopt MLOps/LLMOps Frameworks**: Establish operational processes to manage AI/ML models throughout their lifecycle, from development to deployment and monitoring, ensuring scalability and reliability.
- **Mechanisms of Monitoring**: Establish mechanisms to track progress and adapt the strategy as needed. Regularly review the strategy's effectiveness in light of technological advancements and changing business needs. Set clear performance benchmarks for AI initiatives to track progress and ensure alignment with business objectives. Regularly review these metrics to adapt strategies as needed.
  - **Continuous Monitoring**: Implement a monitoring system to track AI model performance and ensure they consistently deliver value. Regularly update models to maintain accuracy and align them with changing business needs.
  - **Iterative Improvements**: Use feedback from pilots and early deployments to refine and iterate the AI roadmap. Stay flexible to accommodate new technological advances or shifts in business priorities.

    - **Define Success Metrics**: Establish clear criteria for measuring AI success, such as cost savings, revenue growth, customer satisfaction, or productivity improvements.
    - **Demonstrate Value**: Regularly report to stakeholders on the ROI of AI initiatives, showcasing wins and highlighting areas for future growth.
- **Ethical Considerations and Responsible AI Principles**: Incorporate ethical considerations into the strategy to address issues like bias, transparency, and regulatory compliance. Establish responsible AI principles to guide development and deployment.
    - **Bias and Fairness**: Ensure AI systems are designed to be unbiased and fair, particularly in critical areas like recruitment, healthcare, or finance. Regularly audit the systems for bias.
    - **Data Privacy & Security**: Ensure compliance with data privacy regulations (e.g., GDPR) and implement robust data security measures to protect sensitive information.
    - **Explainability & Transparency**: To foster trust and accountability, build models that are explainable, especially for high-stakes decisions.

By incorporating these components into an AI strategy roadmap, organizations can effectively navigate the complexities of AI adoption, maximize their investment in AI technologies, and drive meaningful business outcomes.

REFERENCES

Accenture. 2020. *AI: Built to Scale – Achieving Growth and Efficiency with AI*. Accenture Strategy.

Accenture. 2021. *From Experimentation to Transformation: How Leading Organizations Scale AI*. Accenture.

AI4 ALL and the World Economic Forum. 2019. *The AI Readiness Framework*. World Economic Forum.

Boston Consulting Group. 2019. *The CEO's Guide to AI Strategy*. Boston Consulting Group.

Boston Consulting Group. 2020. *AI for Business: How to Design an AI Strategy*. Boston Consulting Group.

Chui, M. and Hall, B. 2022. *The State of AI in 2022 – and a Half Decade in Review*. McKinsey Global Institute.

Deloitte. 2020. *AI Adoption in Business: Unlocking Value with AI*. Deloitte Review.

Deloitte. 2021. *Implementing AI Successfully: Strategies for CIOs and Business Leaders*. Deloitte.

Deloitte. 2022. *State of AI in the Enterprise*, 5th Edition. Deloitte Insights.

European Commission. 2020. *White Paper on Artificial Intelligence: A European Approach to Excellence and Trust*. European Union.

Gartner. 2019. *AI Strategy by Design: Aligning AI Initiatives with Business Outcomes*. Gartner.

Gartner. 2020. *How to Build an AI Strategy that Aligns with Business Objectives*. Gartner Research.

Gartner. 2021. *Top Strategic Predictions for 2022 and Beyond Leveraging AI for Business Growth*. Gartner.

McKinsey & Company. 2020. *Building an AI Strategy: Practical Steps for Business Leaders*. McKinsey Digital.

McKinsey & Company. 2019. *How Artificial Intelligence Will Transform Business and Drive Value*. McKinsey.

MIT Sloan Management Review. 2020. *Artificial Intelligence and Business Strategy: An Executive Guide*. MIT Sloan.

OECD, 2019. *OECD AI Principles: Building Trustworthy AI*. Organization for Economic Co-operation and Development.

PwC. 2020. *AI in the Enterprise: From Promise to Performance*. PwC.

PwC. 2021. *AI Predictions 2021: Five AI Trends for 2021*. PwC.

PwC. 2019. *Harnessing the Power of AI: A Strategy Guide for Business Leaders*. PwC.

Ransbotham, S. et al. 2019. *Reshaping Business with Artificial Intelligence*. MIT Sloan Management Review.

Ransbotham, S. et al. 2020. *Winning with AI: Pioneers Combine Strategy, Organizational Muscle, and Tech*. BCG Henderson Institute.

Schwab, K. and Davis, N. 2018. *Shaping the Fourth Industrial Revolution*. World Economic Forum.

World Economic Forum. 2020. *AI for Business: A Framework for Developing an AI Strategy*. World Economic Forum.

World Economic Forum. 2018. *Harnessing Artificial Intelligence for the Earth*. World Economic Forum.

11 Facets of Responsible AI

11.1 INTRODUCTION

AI systems increasingly exert an extensive impact on various facets of our existence, encompassing healthcare, the quality of education we receive, the determination of which news articles or social media posts we encounter, the allocation of employment opportunities, detention decisions, and the intensification of policing efforts in some areas, to name a few. These risks are even more pronounced with the recent advent of Gen AI and the impact these systems have on various societal aspects. Gen AI has democratized the use of AI utilization not only in Businesses but also integrated into people's ways of life. Although people find Gen AI entertaining and engaging, if no precautions are taken, they can enhance discrimination, foster unfair treatment, and even make harmful decisions against certain individuals or populations (Figure 11.1).

Responsible AI (RAI) is making the design, development, and use of AI socially sustainable, reaping the benefits of innovation while controlling the risks. Widespread recognition of the importance of RAI has led to recent legislative and regulatory decisions and high-level directives. Examples include the U.S. Executive Order on Safe, Secure, and Trustworthy AI, which recognizes the potential of AI while committing to address "safe, responsible, and non-discriminatory uses of AI" and the adoption of the AI Act by the European Union.

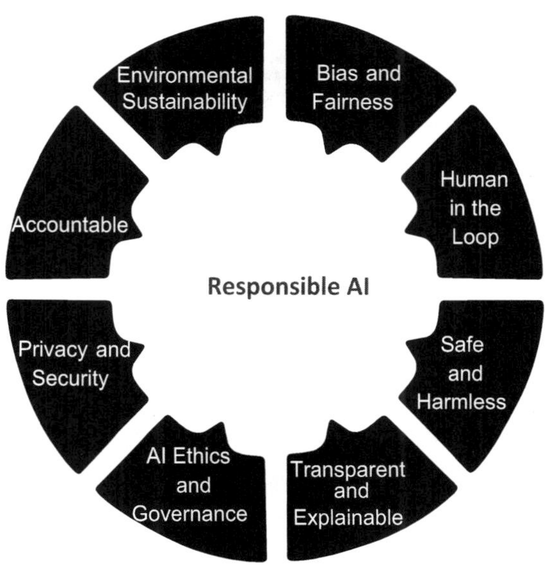

FIGURE 11.1 Dimensions of Responsible AI.

DOI: 10.1201/9781003190325-11

Stakeholders, including technical developers, designers, end users, and other individuals impacted by AI and society at large, all have distinct priorities. For this reason, deep engagement with the tensions between differing stakeholder perspectives is necessary to build and deploy AI systems responsibly. Naturally, industry practitioners play a decisive role in our collective ability to achieve the goals of RAI, and industry partners must work closely with academic institutions to integrate RAI advances into consumer-facing applications and services.

11.2 KEY PRINCIPLES OF RAI

RAI is about creating and utilizing ethical, transparent, just, diverse, and safe AI systems. It ensures that AI systems are created and used based on social values and respect for individuals' rights. RAI is built on several foundational principles that guide the use of AI:

11.2.1 Bias and Fairness

Biases in LLMs generally result from their training data and the algorithms used in their development. These biases can be social, linguistic, or representational biases.

Social biases reflect model biases related to various social statuses and characteristics, such as gender, age, sexual orientation, nationality, ableism, physical appearance, political affiliation, income, race, occupation, education, and lifestyle choices. These biases in LLMs result from the training data on which they are trained. LLMs can perpetuate biases in the training data, reinforcing stereotypes and unequal performance in content generation related to different social groups.

Language biases refer to LLMs' preferential treatment or unequal performance across different languages, dialects, or linguistic styles. These biases can lead to unequal service quality or information accuracy, reinforcing social inequalities in education, communication, and business.

Representation biases in LLMs occur when certain groups, perspectives, or types of information, including demographic, cultural, and socioeconomic representations, are underrepresented or misrepresented in the training data of these models. The consequences of representation bias can reinforce stereotypes, perpetuate inequalities, and fail to serve diverse user groups effectively (Figure 11.2).

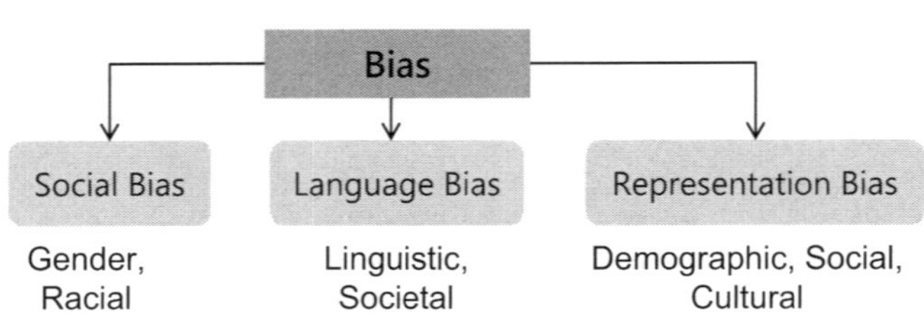

FIGURE 11.2 Biases in LLMs.

Fairness is the quality or state of remaining impartial. It can mean different things to different people and disciplines in various contexts. In Law, fairness includes protecting individuals and groups from discrimination or mistreatment. It focuses on prohibiting behaviors and biases and basing decisions on certain protected factors or social group categories. In Social sciences, fairness is seen in light of social relationships, power dynamics, institutions, and markets. Members of certain groups (or identities) tend to experience advantages. In Quantitative fields such as mathematics, computer science, statistics, and economics, fairness is seen as a mathematical problem. Fairness matches criteria, such as equal or equitable allocation, representation, or error rates, for a particular task or situation. In Philosophy, the ideas of fairness rest on a sense that what is fair is also what is morally right. Political philosophy connects fairness to notions of justice and equity. Fairness is the principle that AI should treat all individuals and groups equitably, ensuring that the system's decisions systematically disadvantage no one. Achieving fairness in AI is not only a technical challenge but also an ethical and societal one.

AI systems tend to learn from data collected and historical decisions made, which can have their own biases based on societal disparities. Training datasets, often historical models, occasionally encompass biases in the model's decision-making. For instance, a face recognition system highly trained on lighter-skinned individuals may struggle to accurately identify healthy individuals from darker races. Another example is that an AI trained on recruitment with a gender skew will likely imitate the bias, favoring certain male candidates for some positions rather than female candidates.

Detecting and preventing these biases is a sensitive process that involves data collection, model training, and testing. AI bias is particularly concerning because it can be challenging to detect and correct, especially in complex AI models. As AI systems become more integrated into decision-making processes, they increase the chances of perpetuating and exacerbating existing structural biases. Thus, it is imperative to comprehend the sources of bias and adopt measures to reduce them to create equitable RAI systems.

Bias in AI systems can be addressed with a combination of technical approaches. Here, we will discuss a technical approach with three stages of the AI pipeline: pre-processing, in-processing, and post-processing (Figure 11.3).

Pre-processing techniques focus on improving the quality of training data before it is used to train the model. This can involve methods like *Data balancing,* where underrepresented groups can be oversampled to ensure equitable representation, and

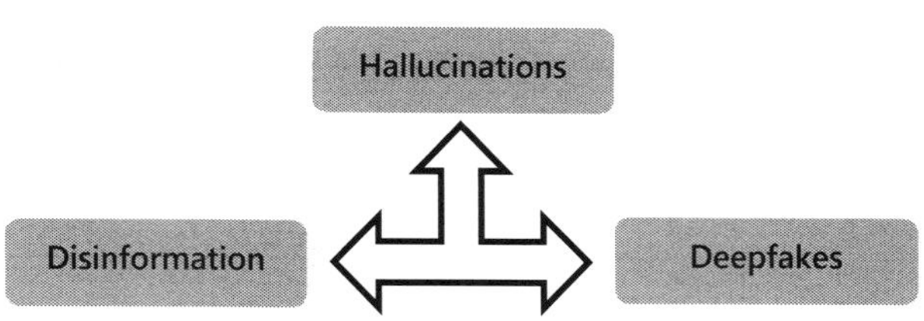

FIGURE 11.3 Trifecta affecting safety and Harm in AI.

Data anonymization, where Customer profiling attributes such as race or gender are removed or obscured to prevent the model from making decisions based on those characteristics.

In-processing techniques modify the algorithm during the training phase to promote fairness. One approach is to use fairness-aware ML algorithms that explicitly account for fairness constraints when optimizing model performance. These algorithms may include fairness metrics, as shown in Table 11.1, in their loss functions, penalizing the model when it exhibits biased behavior. These metrics help AI systems identify, quantify, assess, and reduce bias.

Different applications may require different definitions of fairness, making it a complex issue to address universally. There are multiple ways to define fairness in AI, including *Equal treatment, which ensures that all individuals are treated identically, regardless of their attributes; equal opportunity, which ensures that individuals with similar qualifications or circumstances receive similar outcomes; and Group fairness, which ensures* equitable decisions across demographic groups, such as gender, race, or age.

Post-processing techniques are applied when the model has been trained and typically involve adjusting the model's output to correct for biases. For example, suppose an AI model is used to predict loan approvals. In that case, the decision

TABLE 11.1
Fairness Metrics

Fairness Metric	Description
Equality of Accuracy across groups	Ensure that both privileged and unprivileged groups have equal prediction accuracy.
Disparate Impact	Compare the proportion of individuals receiving a positive outcome in the unprivileged group to the privileged group.
Group Unawareness	A model's decision would be fair if it is unaware of sensitive attributes.
Demographic Parity	Ensure that different demographic groups experience equal proportions of positive outcomes. This measurement ensures that the model prediction is statistically independent of the protected attribute.
Equal Odds	Ensure that privileged and unprivileged groups have equal TPR and FPR.
False Positive Rate (FPR)	Ensure that all groups have the same FPR.
Equal opportunity Or True Positive Rate (TPR)	Ensure that decisions are not based on protected attributes but rather on qualifications and metrics and that all groups have equal access to opportunities.
Positive Predicted Value	Ensure that the chance of success is equal based on a positive prediction.
Negative Predicted Value	The aim is to ensure that the ratio of correctly rejecting samples out of all the samples the model has rejected is the same for each group.
Predictive Value Parity	It is satisfied when both PPV and NPV are satisfied.

thresholds for different demographic groups can be adjusted to ensure more equitable outcomes.

While bias mitigation techniques are essential, they may not be foolproof. Each approach has its trade-offs, and addressing bias in one aspect may inadvertently introduce biases in another. Therefore, continuous monitoring and iteration are necessary to ensure AI systems remain fair and unbiased over time.

Auditing AI systems for fairness is critical in ensuring responsible AI deployment. Frameworks like AI Fairness 360 (developed by IBM) provide open-source resources for conducting fairness audits. Fairness audits typically involve:

- **Data Audits**: Review the training datasets for potential biases, such as underrepresenting certain demographic groups or skewed labels that may lead to discriminatory outcomes.
- **Algorithmic Audits**: Evaluating the model's decision-making process to ensure it does not disproportionately favor or disadvantage specific groups. This may include testing the model on diverse datasets to assess its performance across different demographics.
- **Outcome Audits**: Analyzing the AI system's outputs to determine if there are any patterns of unfair treatment.

Besides the technical approaches, it is equally important to encourage team diversity when developing AI systems to promote fairness. Diverse teams have a higher tendency to spot and mitigate any bias that may exist in AI systems. Technology, ethics, and subject matter experts must work concurrently to ensure that the AI built is effective and morally appropriate.

11.2.2 Human in the Loop

AI is increasingly used today to support decision-making in delicate domains, including criminal justice, healthcare, and finance. Given AI systems' growing independence and autonomy, human control should be carefully considered. For an AI to reliably guide actions, we should understand that, by design, AI is driven by algorithms and may not make informed and contextually appropriate decisions.

Humans should actively participate in developing and deploying AI (model training, assessment, or operation). Developers should incorporate human expertise into the AI lifecycle to improve accuracy, dependability, and adaptability. While testing AI systems, they should be subjected to various simulated situations, including unusual or extreme ones (edge cases). The goal is to identify any potential responses from the AI that could trigger unsafe actions. This guardrail ensures that these issues are addressed before real-world deployment.

We should consider making AI systems failsafe. We should be careful, particularly when making decisions impacting someone's life, such as identifying a person's illness or approving a loan, and ensuring that these decisions have appropriate human interventions and reviews is essential. AI systems should assist and augment human decision-making by providing insights and recommendations. Safety is pivotal when deploying AI systems. We should remember that these systems are probabilistic black boxes, especially when these agents start making decisions that cross the boundary from digital to physical.

While AI is in action, some fail-safes could include human overrides or programmed instructions to automatically shut down certain processes to prevent the system from causing harm or damage. AI should enhance human well-being by incorporating appropriate checks and balances and help maintain sufficient human control rather than replace or undermine human decision-making. Developing and deploying AI should prioritize human welfare and ensure that the technology complements human intelligence rather than diminishing it.

11.2.3 Safe and Harmless

The advent of AI technologies, especially Gen AI, has created deepfakes and synthetic media, which can distribute false information and influence social perception. These applications can produce materials that are deceptively similar to the truth but aim to harm more than help inform the audience and threaten trust in any information channel. This is the single most important issue that will affect Responsible AI (Figure 11.4).

The RAI for the safe and harmless implementation of AI systems requires a multi-pronged approach. Some technical approaches being researched are geared toward building robust AI verification systems to detect and counteract hallucinations, misinformation, and deepfakes. On the policy and regulation side, all stakeholders must take and enforce regulations to ensure that RAI frameworks are adopted. Ultimately, end-user awareness and education to critically evaluate content and recognize AI-generated manipulation will also be effective. The trifecta of Hallucinations, Dis/misinformation, and Deepfakes is a significant threat to the way we live, work, and operate!

11.2.3.1 Hallucinations

The term "hallucination" originated in psychology. It indicates a detachment from reality and a sense of delusion. In AI, it is referred to as hallucination when LLM models produce results inconsistent with actual data or diverge from reality. Fake content can be surprisingly convincing because it frequently follows grammar rules

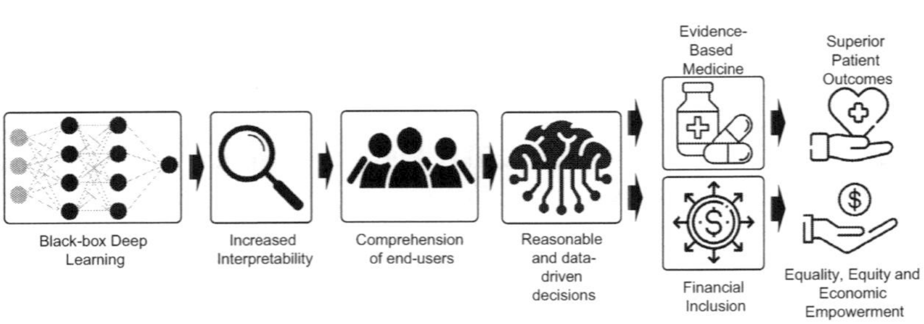

FIGURE 11.4 Use-cases of Interpretability in Financial Services and Medicine.

or has features that sound plausible. Real-world effects have resulted from this, such as attorneys being caught quoting outdated legal precedents. Hallucinations in LLMs are a severe problem, particularly in professions like finance, engineering, medicine, and law, where precision and factual accuracy are critical.

Researchers have extensively studied the causes of hallucinations in LLMs. The primary factor is the quality and content of the training corpora. The problem arises because LLMs are trained on massive datasets of text and code, and they learn to identify patterns and statistical relationships within that data. However, they need to inherently understand the real world or possess the ability to discern truth from fiction. This can lead the model to create seemingly factual, entirely made-up responses. Suppose corpora contain inaccurate or false information and lack sufficient coverage of long-tail knowledge. In that case, it can lead to the generation of hallucinated outputs. The secondary factor lies in the architectural design of LLMs. Despite the architecture of LLMs proving to be capable, in some instances, unidirectional information flow and anomalies within the attention mechanisms can contribute to hallucinations.

Hallucination detection is crucial in assessing the veracity of outputs generated by LLMs. One approach is to introduce external resources to aid detection by retrieving evidence from web sources and authoritative knowledge bases. This evidence serves as a benchmark to verify the veracity of LLM outputs.

Recently, a new form of hallucination called sycophancy has drawn research attention. Sycophancy refers to undesired behavior in which models prioritize agreeing with the user's subjective preferences over providing truthful and objective responses. This preference may include personal opinions, biases, and even factual mistakes expressed by users.

Hallucinations represent a serious risk because they undermine confidence in LLMs. The model loses credibility if a user cannot be certain that the information they receive is true. Hallucinations can also be malicious, as they could be used to spread misinformation or create fake content that sways public opinion.

11.2.3.2 Disinformation

By using various prompts on LLMs, one can create deceptive text, which malicious users can exploit. Disinformation consists of deliberately fabricated information intended to cause harm. Earlier work by malicious users produced disinformation that relied on straightforward prompts. Still, these methods often failed to deceive automated detectors due to their lack of detail and consistency.

Three primary methods have been developed to create more credible fabrications: The standard method refers to embellishing Standard, Mixture, and Chain-of-Thought (CoT). The standard method involves embellishing raw fake information with more convincing details. The mixture method introduces true stories and blends fake content to enhance plausibility. The (CoT) approach involves prompting LLMs with a sequence of operations, instructing them to generate disinformation step-by-step. This CoT technique mirrors human cognitive processes in crafting text that appears more natural and legitimate.

Disinformation may also arise in creating fabricated text when malicious users integrate external knowledge from healthcare and biomedical domains. This domain knowledge might include obsolete materials, inaccurate details, or even malicious content, which can be incorporated to generate unreliable outputs.

Various research efforts have been undertaken to detect and counteract the harmful effects of disinformation. Initial detection models leverage auxiliary information beyond the text of the articles, such as metadata, credibility checks against web sources, emotional and semantic traits, and social media reactions.

11.2.3.3 Deepfakes

The rise of Deepfake, a combination of Gen AI and "fake" media, is a digital fabrication in which realistic likenesses of things are synthetically generated or entirely altered to say or do something that never occurred. Due to the public accessibility of sophisticated Gen AI tools, anyone can craft deepfake content. As these capabilities become democratized, the potential for misuse scales exponentially. Multimedia content (e.g., texts, images, audio, and videos) produced by the LM-based Gen AI tools can now fabricate reality, making it increasingly challenging to discover truth from fiction.

The implications of Gen AI technologies are profound and multifaceted. Countries worldwide grapple with the ramifications of Gen AI content regarding electoral processes and public opinion, while netizens face unprecedented threats to their privacy and security (as the deepfakes created from their public data may act without their consent or even knowledge). Furthermore, the media landscape, the traditional outlet for of factual information propagation, is undergoing a seismic shift as journalists and content creators confront the existential question of what trustworthy information is in the post-deepfake era.

The societal implications of deepfakes and mis/disinformation generated by Gen AI will have unprecedented impacts, touching upon every facet of modern life—from politics and security to individual rights and societal trust. They affect Democracy and Public Opinion, Privacy and Personal Security, Consequences for Media and Journalism, Erosion of Public Trust, and Legal and Ethical Dilemmas.

Addressing this challenge requires combining technical solutions, such as detection algorithms and policy interventions, to regulate the misuse of AI-generated content. The defense against these threats is an active field of research. This book does not discuss current research in this area.

11.2.4 Transparency and Explainability

Transparency in AI nurtures a conducive environment for adopting AI tools into user's workflows. It's necessary to provide Transparency, especially in cases where AI outputs could affect people's lives. For example, in the case of AI in medicine, patients and physicians need to know how such systems arrive at a diagnosis or decide on the treatment of a condition. Financial institutions using AI to approve loans or credit must ensure that their decisions are transparent to regulators and customers. Transparency must comply with legal obligations in highly regulated industries like Financial Services.

Explainability is important regarding the relationship between AIs and humans and their trust in each other. The more opaque or inscrutable the AI system's

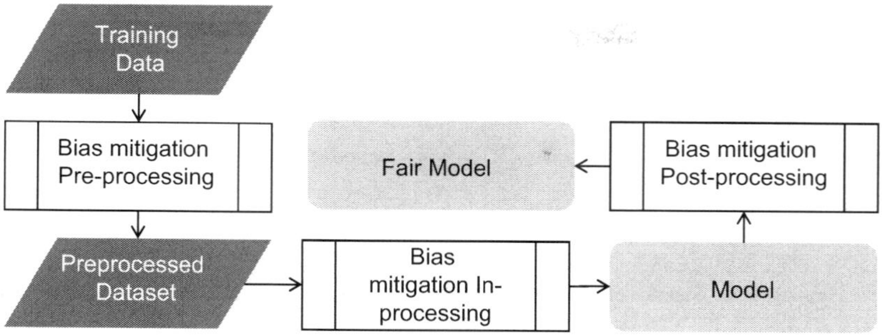

FIGURE 11.5 Bias mitigation processing.

decisions, the less likely the user will trust the system. Explainable AI tools enable AI practitioners to understand the reasons behind a model's decisions, facilitating the identification of any potential bias and improving the model's equity. Figure 11.5 illustrates two scenarios in which AI can drastically improve an AI system's outcomes.

Transparency is also about the openness with which AI systems operate, including the processes involved in model development, data usage, and decision-making. It goes beyond just making the models interpretable; it also includes disclosing the data sources and rationale behind the model design at every stage of the AI pipeline. The system's intuitive design becomes imperative for users or other stakeholders to understand how an AI system arrives at a conclusion or action.

One obstacle is the complexity of AI systems. Deep learning outputs are considered a "black box," and explaining how such a model arrived at its conclusions can be challenging. For instance, LLMs contain billions of parameters, making it extremely difficult to understand exactly how they generate a particular text. The lack of Explainability and interpretability makes it hard to trust an LLM's output, especially in high-stakes situations.

Interpretability refers to the ability to explain or provide meaning in a way humans can understand. Explainability is the ability to give meaning, whereas interpretability is the ability to understand. Explainability involves providing details or reasons to facilitate or clarify the understanding of AI systems. An interpretable AI system is transparent about the relationships between the inputs and outputs. Highly interpretable models are simpler but may lack the predictive accuracy of more complex models. To ensure model "Explainability" and "Transparency," self-explainable models can be developed, and their decision paths can be visually explained via attribution maps and counterfactuals.

Achieving RAI through transparency requires diverse stakeholders, from AI specialists to potential end-users, to collaborate while building AI systems that make AI an accessible, understandable, and trusted resource.

In the case of AI, Explainability means the system accepts, understands, and explains the mechanisms responsible for its actions. Explainability as a concept is being employed by stakeholders such as users and regulators as an area of concern regarding how AI makes system decisions or provides recommendations.

There are two approaches to enhancing Explainability in AI systems:

- **Intrinsic Explainability**: This strategy uses models that can be understood automatically, such as decision trees, linear regression, or rule-based frameworks. These choices are made since the decision-making process is, by design, easy to follow.
- **Post-hoc Explainability**: For more complex, black-box models like deep learning or ensemble methods, post-hoc explainability methods include:
  - **Feature Importance**: This method ranks the input features based on their contribution to the model's predictions. For example, a credit scoring model might show that income, age, and employment status are the most influential factors in determining whether an individual receives a loan.
  - **LIME (Local Interpretable Model-Agnostic Explanations)**: This method creates simple, interpretable models around individual predictions. It explains why the features make a particular decision for a specific input by approximating its behavior with a "linear" local model.
  - **SHAP (SHapley Additive exPlanations)**: SHAP values consistently calculate feature importance based on cooperative game theory. SHAP explains which features contributed to a decision and quantifies the direction and magnitude of each feature's contribution.
  - **Counterfactual Explanations**: Counterfactual explanations focus on "what-if" scenarios, showing how small changes to the input data could have led to different outcomes. For example, suppose a customer is denied a loan. In that case, a counterfactual explanation might indicate that increasing their income by $5,000 would have resulted in approval.

These explainability techniques are crucial for translating complex model behavior into understandable insights for stakeholders. Techniques like LIME and SHAP are helping bridge this gap by making even complex models more interpretable. As these techniques evolve, the industry may achieve a better balance between model complexity and Transparency, enabling organizations to deploy high-performing AI systems.

11.2.5 Ethics and AI Governance

Ethics is a set of moral principles that help us discern right from wrong. The study of AI ethics is a multidisciplinary field that aims to minimize risks and negative effects while maximizing AI's positive effects. Ethics is not a one-size-fits-all approach, as different cultures and societies have varying values and ethical standards. What may be considered ethical in one country may not be viewed the same way in another. This creates a challenge for multinational organizations that operate across borders. Balancing global ethical standards with local cultural sensitivities requires careful consideration and collaboration between governments, organizations, and civil society.

With the emergence of Big Data, companies have increased their focus on driving automation and data-driven decision-making across their organizations. While the goal is to improve business outcomes, companies are experiencing unforeseen challenges in some of their AI applications. Ethical standards must direct the development and application of AI systems, since they have the potential to exacerbate preexisting biases in fields like criminal justice, healthcare (medical diagnosis), finance (loan approvals), policing (law enforcement), and employment. Certain AI applications disproportionately impact marginalized populations, and these tangible implications demonstrate the urgency of formulating and adhering to responsible AI policies and frameworks to ensure the deployment of equitable, fair, and safe AI systems.

As instances of unfair outcomes have come to light, some leading tech organizations, Analyst firms, and management consulting companies are forming new guidelines and frameworks to address concerns about the ethics of AI.

As with all technical advancements, innovation typically outpaces government regulation in emerging sectors. As governments gain a deeper understanding, businesses can expect additional AI guidelines to adhere to, allowing them to prevent violations of civil liberties and human rights. AI technologies must be designed and adopted with a strong focus on human dignity and morality. Ethical AI ensures that systems do no harm, avoid manipulation, and prioritize the well-being of humanity.

AI governance ensures that AI systems function within a clear, ethical framework. It describes the procedures, guidelines, and standards that support the ethical and secure application of AI systems. To support and guarantee security, equity, and human rights, AI governance frameworks guide AI research, development, and use. Governance is not just about overseeing AI models but also including the data used for model development. It includes effective Data governance practices, setting ethical and integrity standards for data acquisition, storage, and utilization. Effective AI governance includes oversight mechanisms that address risks such as bias, privacy infringement, and misuse, while nurturing innovation and building trust. Successful AI governance necessitates the participation of stakeholders such as researchers, developers, users, managers, and ethicists.

The governance of AI systems is a multifaceted challenge that operates on multiple levels, ranging from research groups to overarching governmental frameworks. While each level plays a critical role in ensuring the development and deployment of RAI, delineating responsibilities and activation points for different governance layers remains unclear. At the research group level, the focus is often on ethical considerations and compliance with institutional policies. Institutions should establish guidelines that strike a balance between innovation and social responsibility. However, the role of government in AI governance introduces an additional layer of complexity. Determining when and how government intervention should occur and the form it should take remains a topic of debate.

11.2.6 Security and Privacy

Protecting individual privacy is a fundamental ethical concern in AI. AI systems often rely on large amounts of personal data, raising important questions about data ownership, consent, and protection. RAI systems must implement robust safeguards to prevent unauthorized access, misuse of personal data, and potential breaches. The primary issue is data leakage, where personal information included in training datasets can be exposed during model interactions, particularly when using LLMs. This concern is closely tied to the challenge of re-identifying anonymized data, where seemingly non-identifying information can be recombined by the model to reveal individual identities.

Data protection is paramount in AI, especially considering the sensitive nature of the data involved. Most datasets include personally identifiable information like medical histories, credit card details, social platform usage, fingerprints, and voice recognition audios. We all know we could use the most private data in models trained, for example, on social networks containing user-exposed profiles. This issue becomes more complicated when AI systems are integrated into architecture within sensitive fields such as healthcare, finance, or policing. In this context, anonymization, encryption, and data minimization are essential strategies to reduce privacy risks in AI. Regulations like the General Data Protection Regulation (GDPR) and the California Consumer Privacy Act (CCPA) are key in controlling how personal data is collected and used.

Security risks in AI systems involve vulnerabilities that malicious actors can exploit to manipulate models or steal sensitive information. As AI systems become more pervasive, they increasingly become targets of cyberattacks.

One of AI's most prominent security threats is the risk of adversarial attacks. In adversarial attacks, small, imperceptible changes are made to input data, which can trick an AI model into making incorrect predictions or classifications. For example, an adversarial attack on a facial recognition system might manipulate an image so that the system misidentifies the person, creating security loopholes.

Another significant security risk is model poisoning. In this type of attack, an adversary introduces malicious data into the training set, causing the model to learn harmful behaviors or make biased decisions. If undetected, model poisoning can undermine the reliability and safety of AI systems.

Additionally, the confidentiality of AI models is also at risk. Sophisticated attackers can reverse-engineer AI models or extract sensitive information from them. For example, in individual interference attacks, AI models can infer whether a particular individual's data is part of the model's training. This attack threatens the privacy of users whose data contributed to the model's development.

To address privacy concerns, organizations can adopt privacy-preserving methods such as:

- **Federated Learning**: Conventional AI systems gather data from different sources to a commonplace and build models based on that data. In contrast, Federated learning enables training AI models on the devices and locations where the data is generated, for instance, edge devices or mobile phones.

Subsequently, these local models are combined to create a global model without transmitting data outside the device. This approach effectively mitigates the threats to the user's privacy by keeping the sensitive information and its potential exposure distributed.

- **Differential Models**: Differential privacy adds statistical noise to the data or the model's outputs, ensuring that individual data points cannot be easily identified while allowing the AI system to make accurate predictions on a broader population level. This technique is widely used to balance the utility of AI models with the need for data privacy.
- **Encrypted Computation**: Techniques like homomorphic encryption allow AI models to perform computations on encrypted data without decrypting it. This ensures that sensitive information remains protected throughout the entire AI process.
- **Synthetic Data**: Synthetic data involves generating artificial datasets that replicate the statistical properties of real data without containing any personally identifiable information. These datasets can be used to train AI models, minimizing the need for actual sensitive data while still achieving accurate results.

Building a secure AI pipeline involves safeguarding the entire AI lifecycle, from data collection to model deployment and monitoring. This requires implementing security best practices such as:

- **Data Encryption**: The process of encoding data when stored and in motion so that it is not accessed or altered by anyone unauthorized.
- **Access Controls**: Implementing role-based access control for sensitive information and AI models ensures that unauthorized entities cannot compromise any key elements in the AI pipeline.
- **Regular Audits and Testing**: Regular security and vulnerability assessments help identify potential weaknesses in the AI system before they can be exploited. Penetration defense red-teaming exercises can simulate attacks and expose flaws in system defenses.
- **Monitoring and Anomaly Detection**: allows organizations to identify and respond to suspicious activities that could indicate a security breach or adversarial attack.

Privacy and security are fundamental components of RAI. To reduce security risks, institutions need to use multiple technologies to secure the data pipeline, ensure the model's safety during both training and operational phases, and monitor for any unusual behavior or activities in AI.

11.2.7 Accountability

Accountability is necessary to guarantee auditable decision-making procedures. It is essential for maintaining public confidence and fulfilling the rising standards set by oversight organizations. Gen AI tends to generate false or misleading information by amplifying biases in its training data. The development of counterfactual speech makes the goal of responsible AI much more challenging. For instance, creating false legal cases in the legal field and falsifying references in scientific article writing highlight the dangers of applying Gen AI in crucial fields.

Everybody concerned with the evolution and deployment of AI systems, such as developers, companies, and even stakeholders, is accountable for the systems they create. In the unfortunate situation where an AI system misbehaves or even causes harm, a system must be in place to address such a situation. This brings in the concept of having oversight roles and seeking corrective actions to ensure accountability of people, processes, or systems when things go wrong.

11.2.8 Environmental Sustainability

AI has the potential to impact the environment significantly. AI models require substantial energy and natural resources during training, deployment, and inference. For some Gen AI services, such as text and image generation, the total energy consumption for inference can be comparable to, or even exceed, that for training, resulting in substantial carbon emissions and freshwater usage.

Some efforts have been made to enhance efficiency and reduce the energy consumption of AI models. Example strategies include model compression and geographical load balancing (GLB). On the infrastructure side, there has been a rise in the adoption of carbon-free energy and climate-conscious cooling system designs in the data center industry. Therefore, employing green energy, enhanced data processing, and implementing energy-saving algorithms is essential.

11.3 GUIDELINES AND FRAMEWORKS FOR RAI

Frameworks typically focus on several dimensions, such as AI governance, risk management, and ethical and legal requirements of Compliance. These frameworks can also be adapted to specific industries, helping businesses develop sector-specific strategies to manage AI risks. By adopting these frameworks, organizations can standardize their approach to RAI. Various organizations, governments, and industry groups have guidelines and frameworks to promote RAI:

- **The European Commission's Ethical Guidelines for Trustworthy AI**: These guidelines highlight principles like human agency and oversight, technical robustness, Transparency, and diversity. They provide a framework for developing and deploying AI systems that are ethical, legal, and aligned with fundamental rights.
- **IEEE's Ethically Aligned Design** offers a comprehensive framework that promotes human-centered AI and addresses global ethical challenges in AI design and development. Each framework serves as a blueprint for operationalizing RAI across industries, encouraging organizations to adopt ethical AI development practices.
- **The Organization for Economic Co-operation and Development (OECD) AI Principles**: This framework helps ensure the ethical use of AI

in policy and practice. It focuses on Singapore's Transparency and the safety of AI systems. It can be used in regulatory Compliance and organizational policy.

- **NIST AI Risk Management Framework (NIST AI RMF)**: This framework identifies, assesses, and manages risks related to AI systems.
- **Singapore Model AI Governance**: This framework by Singapore's Personal Data Protection Commission (PDPC) provides a practical guide for organizations deploying AI systems. It emphasizes user-centricity and Explainability and helps build an operationally feasible governance structure for large-scale AI deployment.
- **Fairness, Accountability, and Transparency in Machine Learning (FAT/ML)**: This framework promotes fairness and accountability in AI systems, focusing on data biases and model transparency. It also emphasizes algorithmic fairness and avoids bias in decision-making systems.
- **Governance of AI (ISO/IEC 38507)**: This ISO standard guides how boards and executives can oversee and govern AI. This framework helps manage AAI4People's from a top-down perspective, aligning AI with corporate governance principles.
- **The AI Ethics Impact Group Framework**: This broad framework integrates various aspects of AI responsibility, including ethical risks, social acceptance, and regulatory Compliance. It helps develop ethical guardrails that consider societal and cultural impacts.
- **AI4People's Ethical Framework**: This multi-stakeholder initiative sets out four principles (beneficence, non-maleficence, autonomy, and justice) and 20 action points to guide the ethical development of AI.
- **The Partnership on AI**: An organization founded by major technology companies like Google, Amazon, and Microsoft, the Partnership on AI promotes responsible AI development by sharing best practices, research, and guidelines on ethical AI use.
- **The Asilomar AI Principles**: These principles, created by the Future of Life Institute, focus on long-term AI safety and ethics, emphasizing human control, non-maleficence (do no harm), and shared benefits from AI advances.

11.3.1 Toolkits in Operationalizing Responsible AI

Toolkits often provide extensive documentation, tutorials, and guidelines that help implement RAI practices efficiently. RAI toolkits are crucial in operationalizing frameworks' ethical principles and guidelines. Toolkits are collections of software, algorithms, and methodologies designed to help organizations build, assess, and monitor AI systems in line with RAI principles.

- Microsoft's toolkit and Google's What-If Tool allow developers to analyze and visualize the performance of AI models, assess fairness, and evaluate how changes in input data impact model predictions. These tools helps identify potential biases in models, ensuring that decisions made by AI

systems are equitable. Microsoft's Fairlearn helps developers measure the fairness of AI models, make trade-offs between fairness and other performance metrics, and implement bias mitigation strategies.
- IBM's AI Fairness 360 and AI Explainability 360 toolkits provide frameworks for promoting fairness and Explainability in AI models. These tools offer algorithms to detect and mitigate bias and techniques to improve the interpretability of model outputs. They empower developers to build AI systems that are not only accurate but also transparent.
- **PWC RAI Toolkit**: PWC has developed an RAI Toolkit that provides resources for deploying AI with accountability, interpretability, robustness, fairness, and privacy. This toolkit can monitor AI deployments and ensure they adhere to responsible AI principles.

The application of RAI in practice entails the adoption of the frameworks mentioned above and toolkits. Such frameworks and toolkits should be assessed and incorporated into the organizations' prevailing ethical and governance approaches. They should also adjust these frameworks to accommodate regulatory requirements, industry best practices, or organizational policies. To succeed in implementing RAI, it is vital to involve cross-functional teams comprised of AI developers, data scientists, ethicists, lawyers, and business leaders.

REFERENCES

Artificial intelligence: Ethical, social and security impacts for the present and the future. https://dokumen.pub/artificial-intelligence-ethical-social-and-security-impacts-for-the-present-and-the-future-1787783707-9781787783706.html.

Ayyamperumal, S. G. and Ge, L. 2024. Current state of LLM risks and AI guardrails. https://arxiv.org/abs/2406.12934.

Biasness in AI (Artificial Intelligence) - GRM Institute. https://grm.institute/blog/biasness-in-ai-artificial-intelligence/.

Dehghani, F., Dibaji, M., Anzum, F., Dey, L., Basdemir, A., Bayat, S., Boucher, J., Drew, S., Eaton, S. E., Frayne, R., Ginde, G., Harris, A., Ioannou, Y., Lebel, C., Lysack, J., Arzuaga, L. S., Stanley, E., Souza, R., Santos, R. D. and Bento, M. 2024. Trustworthy and responsible AI for human-centric autonomous decision-making systems. https://arxiv.org/abs/2408.15550.

Dignum, V. 2019. *Responsible Artificial Intelligence: How to Develop and Use AI in a Responsible Way*. Cham: Springer.

Floridi, L., Cowls, J., Beltrametti, M., Chatila, R., Chazerand, P., Dignum, V., Luetge, C., Madelin, R., Pagallo, U., Rossi, F. and Schafer, B. 2018. AI4People—An ethical framework for a good AI society: Opportunities, risks, principles, and recommendations. *Minds & Machines* 28: 689–707. doi:10.1007/s11023-018-9482-5.

Goellner, S. and Brumen, B. 2024. Responsible artificial intelligence: A structured literature review. https://arxiv.org/abs/2403.06910.

Jiao, J., Afroogh, S., Xu, Y. and Phillips, C. 2024. Navigating LLM ethics: Advancements, challenges, and future directions. https://arxiv.org/abs/2406.18841.

Lu, Q., Zhu, L., Whittle, J. and Xu, X. 2023. *Responsible AI: Best Practices for Creating Trustworthy AI Systems*, 1st edition. CSIRO.

Lyons, J. B., Hobbs, K., Rogers, S. and Clouse, S. H. 2023. Responsible (use of) AI. *Frontiers in Neuroergonomics* 4. doi:10.3389/fnrgo.2023.1201777.

Mitra, A., Mohanty, S. P. and Kougianos, E. 2024. The world of generative AI: Deepfakes and large language models. https://arxiv.org/abs/2402.04373.

Reddy, S. 2022. Explainability and artificial intelligence in medicine. *The Lancet Digital Health* 4(4): e214–e215.

Sadek, M., Kallina, E., Bohné, T., Mougenot, C., Calvo, R.A. and Cave, S. 2024. Challenges of responsible AI in practice: Scoping review and recommended actions. *AI & SOCIETY*. doi:10.1007/s00146-024-01880-9.

Shoaib, M. R., Wang, Z., Ahvanooey, M. T. and Zhao, J. 2023. Deepfakes, misinformation, and disinformation in the era of frontier AI, generative AI, and large AI models. https://arxiv.org/abs/2311.17394.

Tahaei, M., Constantinides, M., Quercia, D. and Muller, M. 2023. A systematic literature review of human-centered, ethical, and responsible AI. https://arxiv.org/abs/2302.05284.

Appendix I

MEAN SQUARE DEVIATION, S/N RATIO AND ROBUST QUALITY INDEX

MSD	SNR	RQI	MSD	SNR	RQI
100	−20	0	82	−19.14	4.31
99.5	−19.98	0.11	81.5	−19.11	4.44
99	−19.96	0.22	81	−19.08	4.58
98.5	−19.93	0.33	80.5	−19.06	4.71
98	−19.91	0.44	80	−19.03	4.85
97.5	−19.89	0.55	79.5	−19.00	4.98
97	−19.87	0.66	79	−18.98	5.12
96.5	−19.85	0.77	78.5	−18.95	5.26
96	−19.82	0.89	78	−18.92	5.40
95.5	−19.80	1.00	77.5	−18.89	5.53
95	−19.78	1.11	77	−18.86	5.68
94.5	−19.75	1.23	76.5	−18.84	5.82
94	−19.73	1.34	76	−18.81	5.96
93.5	−19.71	1.46	75.5	−18.78	6.10
93	−19.68	1.58	75	−18.75	6.25
92.5	−19.66	1.69	74.5	−18.72	6.39
92	−19.64	1.81	74	−18.69	6.54
91.5	−19.61	1.93	73.5	−18.66	6.69
91	−19.59	2.05	73	−18.63	6.83
90.5	−19.57	2.17	72.5	−18.60	6.98
90	−19.54	2.29	72	−18.57	7.13
89.5	−19.52	2.41	71.5	−18.54	7.28
89	−19.49	2.53	71	−18.51	7.44
88.5	−19.47	2.65	70.5	−18.48	7.59
88	−19.44	2.78	70	−18.45	7.75
87.5	−19.42	2.90	69.5	−18.42	7.90
87	−19.40	3.02	69	−18.39	8.06
86.5	−19.37	3.15	68.5	−18.36	8.22
86	−19.34	3.28	68	−18.33	8.37
85.5	−19.32	3.40	67.5	−18.29	8.53
85	−19.29	3.53	67	−18.26	8.70
84.5	−19.27	3.66	66.5	−18.23	8.86
84	−19.24	3.79	66	−18.20	9.02
83.5	−19.22	3.92	65.5	−18.16	9.19
83	−19.19	4.05	65	−18.13	9.35
82.5	−19.16	4.18	64.5	−18.10	9.52

MEAN SQUARE DEVIATION, S/N RATIO AND ROBUST QUALITY INDEX

MSD	SNR	RQI	MSD	SNR	RQI
64	−18.06	9.69	46	−16.63	16.86
63.5	−18.03	9.86	45.5	−16.58	17.10
63	−17.99	10.03	45	−16.53	17.34
62.5	−17.96	10.21	44.5	−16.48	17.58
62	−17.92	10.38	44	−16.43	17.83
61.5	−17.89	10.56	43.5	−16.38	18.08
61	−17.85	10.73	43	−16.33	18.33
60.5	−17.82	10.91	42.5	−16.28	18.58
60	−17.78	11.09	42	−16.23	18.84
59.5	−17.75	11.27	41.5	−16.18	19.10
59	−17.71	11.46	41	−16.13	19.36
58.5	−17.67	11.64	40.5	−16.07	19.63
58	−17.63	11.83	40	−16.02	19.90
57.5	−17.60	12.02	39.5	−15.97	20.17
57	−17.56	12.21	39	−15.91	20.45
56.5	−17.52	12.40	38.5	−15.85	20.73
56	−17.48	12.59	38	−15.80	21.01
55.5	−17.44	12.79	37.5	−15.74	21.30
55	−17.40	12.98	37	−15.68	21.59
54.5	−17.36	13.18	36.5	−15.62	21.89
54	−17.32	13.38	36	−15.56	22.18
53.5	−17.28	13.58	35.5	−15.50	22.49
53	−17.24	13.79	35	−15.44	22.80
52.5	−17.20	13.99	34.5	−15.38	23.11
52	−17.16	14.20	34	−15.31	23.43
51.5	−17.12	14.41	33.5	−15.25	23.75
51	−17.08	14.62	33	−15.19	24.07
50.5	−17.03	14.84	32.5	−15.12	24.41
50	−16.99	15.05	32	−15.05	24.74
49.5	−16.95	15.27	31.5	−14.98	25.08
49	−16.90	15.49	31	−14.91	25.43
48.5	−16.86	15.71	30.5	−14.84	25.79
48	−16.81	15.94	30	−14.77	26.14
47.5	−16.77	16.17	29.5	−14.70	26.51
47	−16.72	16.40	29	−14.62	26.88
46.5	−16.67	16.63	28.5	−14.55	27.26

MEAN SQUARE DEVIATION, S/N RATIO AND ROBUST QUALITY INDEX

MSD	SNR	RQI	MSD	SNR	RQI
28	−14.47	27.64	10	−10.00	50.00
27.5	−14.39	28.03	9.5	−9.78	51.11
27	−14.31	28.43	9	−9.54	52.29
26.5	−14.23	28.84	8.5	−9.29	53.53
26	−14.15	29.25	8	−9.03	54.85
25.5	−14.07	29.67	7.5	−8.75	56.25
25	−13.98	30.10	7	−8.45	57.75
24.5	−13.89	30.54	6.5	−8.13	59.35
24	−13.80	30.99	6	−7.78	61.09
23.5	−13.71	31.45	5.5	−7.40	62.98
23	−13.62	31.91	5	−6.99	65.05
22.5	−13.52	32.39	4.5	−6.53	67.34
22	−13.42	32.88	4	−6.02	69.90
21.5	−13.32	33.38	3.5	−5.44	72.80
21	−13.22	33.89	3	−4.77	76.14
20.5	−13.12	34.41	2.5	−3.98	80.10
20	−13.01	34.95	2	−3.01	84.95
19.5	−12.90	35.50	1.5	−1.76	91.20
19	−12.79	36.06	1	0.00	100.00
18.5	−12.67	36.64			
18	−12.55	37.24			
17.5	−12.43	37.85			
17	−12.30	38.48			
16.5	−12.17	39.13			
16	−12.04	39.79			
15.5	−11.90	40.48			
15	−11.76	41.20			
14.5	−11.61	41.93			
14	−11.46	42.69			
13.5	−11.30	43.48			
13	−11.14	44.30			
12.5	−10.97	45.15			
12	−10.79	46.04			
11.5	−10.61	46.97			
11	−10.41	47.93			
10.5	−10.21	48.94			

Appendix II

Orthogonal Arrays (OAs)

EXAMPLE ORTHOGONAL ARRAY

OAs are denoted by $L_a(b^c)$

L: Latin square
a: number of runs or experiments
b: number of levels of variables
c: number of columns in the array

EXAMPLE OF 2 LEVEL ORTHOGONAL ARRAY

$L_{32}(2^{31})$ Orthogonal Array

Columns/Combinations	1	2	3	4	5	6	7	8	9	10	11	12	13	14	15
1	1	1	1	1	1	1	1	1	1	1	1	1	1	1	1
2	1	1	1	1	1	1	1	2	2	2	2	2	2	2	2
3	1	1	1	2	2	2	2	1	1	1	1	2	2	2	2
4	1	1	1	2	2	2	2	2	2	2	2	1	1	1	1
5	1	2	2	1	1	2	2	1	1	2	2	1	1	2	2
6	1	2	2	1	1	2	2	2	2	1	1	2	2	1	1
7	1	2	2	2	2	1	1	1	1	2	2	2	2	1	1
8	1	2	2	2	2	1	1	2	2	1	1	1	1	2	2
9	2	1	2	1	2	1	2	1	2	1	2	1	2	1	2
10	2	1	2	1	2	1	2	2	1	2	1	2	1	2	1
11	2	1	2	2	1	2	1	1	2	1	2	2	1	2	1
12	2	1	2	2	1	2	1	2	1	2	1	1	2	1	2
13	2	2	1	1	2	2	1	1	2	2	1	1	2	2	1
14	2	2	1	1	2	2	1	2	1	1	2	2	1	1	2
15	2	2	1	2	1	1	2	1	2	2	1	2	1	1	2
16	2	2	1	2	1	1	2	2	1	1	2	1	2	2	1

EXAMPLE OF 3 LEVEL ORTHOGONAL ARRAY

$L_{32}(2^{31})$ Orthogonal Array

Columns/Combinations	1	2	3	4	5	6	7	8	9	10	11	12	13	14	15	16	17	18	19	20	21	22	23	24	25	26	27	28	29	30	31
1	1	1	1	1	1	1	1	1	1	1	1	1	1	1	1	1	1	1	1	1	1	1	1	1	1	1	1	1	1	1	1
2	1	1	1	1	1	1	1	1	1	1	1	1	1	1	1	2	2	2	2	2	2	2	2	2	2	2	2	2	2	2	2
3	1	1	1	1	1	1	1	2	2	2	2	2	2	2	2	1	1	1	1	1	1	1	1	2	2	2	2	2	2	2	2
4	1	1	1	1	1	1	1	2	2	2	2	2	2	2	2	2	2	2	2	2	2	2	2	1	1	1	1	1	1	1	1
5	1	1	1	2	2	2	2	1	1	1	1	2	2	2	2	1	1	1	1	2	2	2	2	1	1	1	1	2	2	2	2
6	1	1	1	2	2	2	2	1	1	1	1	2	2	2	2	2	2	2	2	1	1	1	1	2	2	2	2	1	1	1	1
7	1	1	1	2	2	2	2	2	2	2	2	1	1	1	1	1	1	1	1	2	2	2	2	2	2	2	2	1	1	1	1
8	1	1	1	2	2	2	2	2	2	2	2	1	1	1	1	2	2	2	2	1	1	1	1	1	1	1	1	2	2	2	2
9	1	2	2	1	1	2	2	1	1	2	2	1	1	2	2	1	1	2	2	1	1	2	2	1	1	2	2	1	1	2	2
10	1	2	2	1	1	2	2	1	1	2	2	1	1	2	2	2	2	1	1	2	2	1	1	2	2	1	1	2	2	1	1
11	1	2	2	1	1	2	2	2	2	1	1	2	2	1	1	1	1	2	2	1	1	2	2	2	2	1	1	2	2	1	1
12	1	2	2	1	1	2	2	2	2	1	1	2	2	1	1	2	2	1	1	2	2	1	1	1	1	2	2	1	1	2	2
13	1	2	2	2	2	1	1	1	1	2	2	2	2	1	1	1	1	2	2	2	2	1	1	1	1	2	2	2	2	1	1
14	1	2	2	2	2	1	1	1	1	2	2	2	2	1	1	2	2	1	1	1	1	2	2	2	2	1	1	1	1	2	2
15	1	2	2	2	2	1	1	2	2	1	1	1	1	2	2	1	1	2	2	2	2	1	1	2	2	1	1	1	1	2	2
16	1	2	2	2	2	1	1	2	2	1	1	1	1	2	2	2	2	1	1	1	1	2	2	1	1	2	2	2	2	1	1
17	2	1	2	1	2	1	2	1	2	1	2	1	2	1	2	1	2	1	2	1	2	1	2	1	2	1	2	1	2	1	2
18	2	1	2	1	2	1	2	1	2	1	2	1	2	1	2	2	1	2	1	2	1	2	1	2	1	2	1	2	1	2	1
19	2	1	2	1	2	1	2	2	1	2	1	2	1	2	1	1	2	1	2	1	2	1	2	2	1	2	1	2	1	2	1
20	2	1	2	1	2	1	2	2	1	2	1	2	1	2	1	2	1	2	1	2	1	2	1	1	2	1	2	1	2	1	2

(Continued)

Columns/Combinations	1	2	3	4	5	6	7	8	9	10	11	12	13	14	15	16	17	18	19	20	21	22	23	24	25	26	27	28	29	30	31
21	2	1	2	2	1	2	1	1	2	1	2	2	1	2	1	1	2	1	2	2	1	2	1	1	2	1	2	2	1	2	1
22	2	1	2	2	1	2	1	1	2	1	2	2	1	2	1	2	1	2	1	1	2	1	2	2	1	2	1	1	2	1	**2**
23	2	1	2	2	1	2	1	2	1	2	1	1	2	1	2	1	2	1	2	2	1	2	1	2	1	2	1	1	2	1	2
24	2	1	2	2	1	2	1	2	1	2	1	1	2	1	2	2	1	2	1	1	2	1	2	1	2	1	2	2	1	2	1
25	2	2	1	1	2	2	1	1	2	2	1	1	2	2	1	1	2	2	1	1	2	2	1	1	2	2	1	1	2	2	1
26	2	2	1	1	2	2	1	1	2	2	1	1	2	2	1	2	1	1	2	2	1	1	2	2	1	1	2	2	1	1	2
27	2	2	1	1	2	2	1	2	1	1	2	2	1	1	2	1	2	2	1	1	2	2	1	2	1	1	2	2	1	1	2
28	2	2	1	1	2	2	1	2	1	1	2	2	1	1	2	2	1	1	2	2	1	1	2	1	2	2	1	1	2	2	1
29	2	2	1	2	1	1	2	1	2	2	1	2	1	1	2	1	2	2	1	2	1	1	2	1	2	2	1	2	1	1	2
30	2	2	1	2	1	1	2	1	2	2	1	2	1	1	2	2	1	1	2	1	2	2	1	2	1	1	2	1	2	2	1
31	2	2	1	2	1	1	2	2	1	1	2	1	2	2	1	1	2	2	1	2	1	1	2	2	1	1	2	1	2	2	1
32	2	2	1	2	1	1	2	2	1	1	2	1	2	2	1	2	1	1	2	1	2	2	1	1	2	2	1	2	1	1	2

Index

Note: **Bold** page numbers refer to tables and *italic* page numbers refer to figures.

Printed in the United States
by Baker & Taylor Publisher Services